SpringerBriefs in Electrical and Computer Engineering

W0225706

For further volumes:
http://www.springer.com/series/10059

Santosh Kulkarni · Prathima Agrawal

Analysis of TCP Performance in Data Center Networks

 Springer

Santosh Kulkarni
Electrical and Computer Engineering
Auburn University
Auburn, AL
USA

Prathima Agrawal
Auburn University
Auburn, AL
USA

ISSN 2191-8112   ISSN 2191-8120 (electronic)
ISBN 978-1-4614-7860-7   ISBN 978-1-4614-7861-4 (eBook)
DOI 10.1007/978-1-4614-7861-4
Springer New York Heidelberg Dordrecht London

Library of Congress Control Number: 2013946993

Printed on acid-free paper

Springer is part of Springer Science+Business Media (www.springer.com)

To our parents

Preface

Cloud Computing is a computing paradigm that delivers hosted services over the Web, based on a *'pay-per-use'* approach. This new style of computing, promises to revolutionize the IT industry by making computing available over the Internet, in a fashion similar to other utilities like water, electricity, gas, cable, and telephony.

Growing adoption of Cloud Computing, by the general public as well as the IT industry, is driving service providers into deploying new data centers. Data centers are sites that host tens of thousands of servers. These servers typically communicate with each other over high speed network interconnects. With growing application deployments, data centers utilize a multi-tiered model where several servers work together to service a single client request. As a result, the overall application performance in a data center, largely depends on the efficiency of its underlying communication fabric.

As far as building communication fabric for a data center goes, there are essentially two choices. The first relies on specialized hardware and protocols like Infiniband, Myrinet, or FibreChannel; the second relies on Ethernet based commodity switches that are available off-the-shelf. Cost and compatibility reasons, however, persuade many data center administrators to employ Ethernet as their baseline communication fabric.

Until a few years ago, Ethernet speeds inside data centers averaged around 100 Mbps. However, recent revisions to IEEE 802.3 standards have led to the development of Ethernet networks, that have speeds of 1 and 10 Gbps. The sudden increase in Ethernet's speeds requires proportional scaling of communication protocols that use it, so that applications that are network intensive can ultimately benefit from the increased bandwidth. Although IP is expected to scale well with evolving Ethernet, there are some legitimate concerns about TCP.

TCP is a protocol standard that is mature and has survived the test of time. However, the unique workloads, speed, and scale of modern data centers violate some of the basic assumptions that TCP was originally based upon. As a result, when TCP is utilized in high-bandwidth, low-latency data center environments, we discover new shortcomings in the protocol. One such shortcoming is referred to as the 'Incast' problem.

TCP Incast is a catastrophic collapse in TCP's throughput that occurs in high bandwidth, low latency network environments when multiple senders communicating with a single receiver, collectively send enough data to surpass the buffering abilities of the receiver's Ethernet switch. The problem arises from a subtle interaction between limited Ethernet switch buffer sizes, TCP's loss recovery mechanisms, and the many-to-one synchronized traffic patterns. Unfortunately, such traffic patterns occur frequently in many data center applications and services. Hence, a feasible solution that addresses the Incast problem is urgently needed.

Our objective in this manuscript, is to address TCP's Incast problem by providing transport layer solutions that are both practical and backward compatible. We approach this goal in two steps. First, we derive an analytical model of TCP Incast. Such a model is essential to understand the reasons behind TCP's throughput collapse. The analytical model provides a closed form equation, which can be used to compute throughput at the client for various synchronized workloads. We verify the accuracy of our model against measurements taken from ns-2 simulations. Next, we discuss some solutions that were designed to address TCP Incast at the transport layer. Specifically, we develop transport layer solutions that improve TCP's performance under Incast traffic, by either proactively detecting network congestion through probabilistic retransmission or by dynamically resizing TCP's segments in order to avoid incurring timeout penalty. We evaluate the merits of the aforementioned solutions using ns-2 simulations. Results show that each of our suggested techniques outperforms standard TCP under various experimental conditions.

Auburn, AL, July 2013   Santosh Kulkarni
                                                                                    Prathima Agrawal

Acknowledgments

We would like to thank the Wireless Engineering Research and Education Center (WEREC) at Auburn University, Auburn, AL for helping us fund and support our research. We would also like to thank Springer, for giving us the opportunity to get our study published. This work was supported in part by the US National Science Foundation (NSF) under Grants ECCS-0947832 and IIP-0738088.

Contents

Acronyms

ABTT	Anterior Block Transfer Timeout
ACK	Acknowledgment
API	Application Programming Interface
ATM	Asynchronous Transfer Mode
CE	Congestion Experienced
ECN	Explicit Congestion Notification
FTP	File Transfer Protocol
HTTP	Hypertext Transfer Protocol
IaaS	Infrastructure as a Service
IBTT	Intermediate Block Transfer Timeout
i.i.d.	Independent and identically distributed
IP	Internet Protocol
IT	Information Technology
MSS	Maximum Segment Size
MTU	Maximum Transfer Unit
NASD	Network Attached Secure Disk
NIST	National Institute of Standards and Technology
NNTP	Network News Transfer Protocol
PaaS	Platform as a Service
RTO	Retransmission Timeout
RTT	Round Trip Time
SaaS	Software as a Service
SMTP	Simple Mail Transfer Protocol
SRU	Server Request Unit
SSH	Secure Shell Protocol
TCP	Transmission Control Protocol
TD	Triple Duplicate Acknowledgments
TDP	Period between two consecutive Triple Duplicate ACKs
TO	TCP Timeout
TOR	Top Of Rack
UDP	User Datagram Protocol

Chapter 1
Introduction

Speaking at the MIT Centennial in 1961, Dr. John McCarthy [1], a leading scientist who pioneered the concept of timesharing [2], said: *"If computers of the kind I have advocated become the computers of the future, then computing may someday be organized as a public utility just as the telephone system is a public utility... The computer utility could become the basis of a new and important industry."* Fifty years on, the Information Technology (IT) industry is finally on the brink of realizing Dr. McCarthy's vision for computing utilities.

With significant advances in information and communications technology over the last five decades, computing is now on the verge of becoming the next utility behind water, electricity, gas, cable and telephony. This computing utility promises to provide the user community with a basic level of service that is sufficient to meet their everyday computing needs [3]. To herald this new era of utility computing, a number of computing models have been proposed, of which Cloud Computing is the latest one.

Cloud Computing is a computing paradigm that delivers hosted services over the Web, based on a *'pay-per-use'* approach. It derives its name from the *'cloud'* symbol that is often used to represent the Internet in networking diagrams and promises to revolutionize the computing industry by making IT available over the Internet [4]. However, Cloud Computing is still an evolving paradigm and as yet, there is no single, widely accepted definition for it. Garnter in [5], defines Cloud Computing as a style of computing where a scalable and elastic IT-related capabilities are provided as a service to external customers using Internet technologies. Forrester in [6], suggests that Cloud Computing refers to a pool of abstracted, highly scalable and managed infrastructure capable of hosting end customer applications and billed by consumption. NIST (National Institute of Standards and Technology) in [7], defines Cloud Computing as a computing model for enabling ubiquitous, convenient, on-demand network access to a shared pool of configurable computing resources (e.g., networks, servers, storage, applications, and services) that can be rapidly provisioned and released with minimal management effort or service provider interaction.

S. Kulkarni and P. Agrawal, *Analysis of TCP Performance in Data Center Networks*, SpringerBriefs in Electrical and Computer Engineering, DOI: 10.1007/978-1-4614-7861-4_1, © The Author(s) 2014

Among the numerous definitions for Cloud Computing, the NIST definition is meant to serve as a reference for broad comparisons of hosted services and deployment strategies. The NIST definition is also intended to provide a baseline for discussions ranging from *'What is Cloud?'* to *'How to best use it?'* [7]. Hence, we adopt NIST's definition of Cloud Computing for the remainder of this document.

1.1 NIST's Model of Cloud Computing

In accordance to the definition from NIST, Cloud Computing covers more than just computing technology. As shown in a three-dimensional diagram in Fig. 1.1 from [8], the model of Cloud Computing is actually composed of five essential characteristics, four deployment models and three service models.

In the subsections below, we outline the key characteristics of Cloud Computing along with a brief overview on the service models and the deployment approaches that are associated with it.

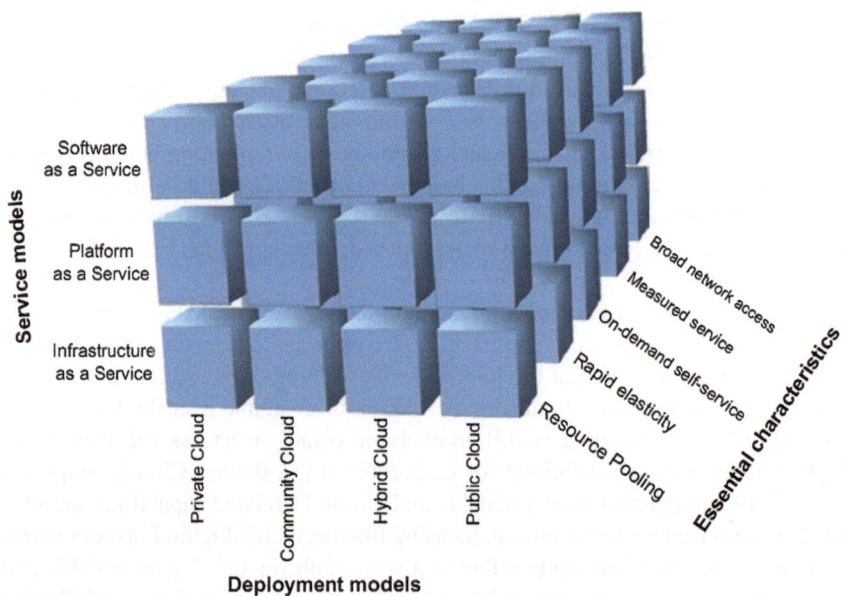

Fig. 1.1 NIST's model of cloud computing

1.1.1 Cloud Characteristics

According to NIST in [7], the essential characteristics of Cloud environment include:

- *On-demand self service* that enables consumers to unilaterally provision computing capabilities, such network storage and server time as needed, automatically, without requiring human involvement.
- *Broad network access* which ensures that all Cloud functionalities and the resources are available over the network and can be accessed through standard mechanisms via thick or thin clients (e.g., laptops, desktops, tablets and mobile phones).
- *Resource pooling* which allows the computing resources provisioned by the provider to be pooled, in order to serve numerous consumers using a multi-tenant model, where different physical and virtual resources are dynamically assigned and reassigned according to the demands of the consumer.
- *Rapid elasticity and scaling* that not only allows the functionalities and resources to scale rapidly outward and inward in accordance to the demands of the consumer, but also allows those capabilities to be elastically provisioned and released.
- *Measured service* that facilitates automatic control and optimization of resource allocations in addition to providing the capability to monitor, control and report resource usage, for both the providers as well as the consumers.

1.1.2 Cloud Service Models

In NIST's model of Cloud Computing, providers offer their services in three flavors, namely, Infrastructure as a Service (IaaS), Platform as a Service (PaaS) and Software as a Service (SaaS) [7]. Among the three service models, IaaS offers the most basic form of Cloud Computing. The three service models can be represented as a pyramid, as depicted in Fig. 1.2, where SaaS is at the top and IaaS is at the bottom. Abstraction among the service models increases as we move towards the top of the pyramid in Fig. 1.2, while the element of control among the service models increases as we move towards the bottom of the pyramid.

- *Software as a Service*—SaaS refers to software applications that are deployed as a hosted services on the Cloud infrastructure. Consumers typically access these applications from client devices that support thin client interfaces like a web browser or use Application Programming Interfaces (API) defined by the hosted software. Under the SaaS service model, consumers do not control or manage the underlying infrastructure or platform. Their only control is usually limited to user specific application configuration settings. Examples of SaaS include: Gmail, Google Docs, Salesforce.com and Microsoft Office 365.
- *Platform as a Service*—PaaS refers to the service where, the providers deliver a computing platform using which consumers can build and deploy their own applications on the Cloud. The computing platform delivered typically includes

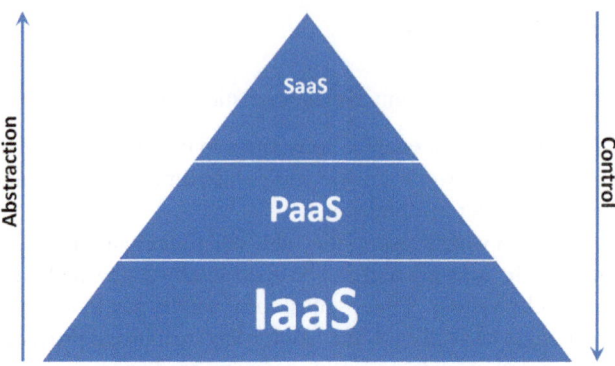

Fig. 1.2 Pyramid of service models in cloud computing

operating systems, compilers, programming libraries and tools that are supported by the service provider. Under the PaaS service model, subscribers do not have access to the underlying Cloud infrastructure. However, they are typically able to control the deployed applications and configuration settings for the application-hosting environment. Examples of PaaS include: Google App Engine, Microsoft Azure and Amazon Elastic Beanstalk.

- *Infrastructure as a Service*—IaaS delivers compute services, typically in the form of a set of virtual machines with associated storage, processing capability, other relevant resources like network connectivity [4]. Under this model, consumers are given the capability to provision computing resources that are made available by service providers. Consumers also have the capacity to deploy and run arbitrary software including operating systems and other applications on the provisioned resources. However, consumers do not have access to the underlying Cloud infrastructure. Their control is limited to operating systems, storage and applications that are deployed by them. Some examples of IaaS include: Amazon CloudFormation, Rackspace Cloud and Google Compute Engine.

1.1.3 Cloud Deployment Models

Cloud deployment approaches represent the way providers deploy Cloud service models in order to make Cloud functionalities available to their consumers. Organizations choose Cloud deployment models based on their specific business, operational and technical requirements [4]. As depicted in Fig. 1.1, NIST categorizes Clouds deployments as Public, Private, Community or Hybrid [7].

- *Public Clouds*—Under Public deployment model, the Cloud functionalities and resources are made available for open use to the general public. Customers access and use hosted Cloud services that are either free or offered on *pay-per-use* basis.

Generally, public Cloud service providers like Microsoft, Amazon and Google own and operate their Cloud infrastructure and offer access to end users via the Internet.

- *Private Clouds*—Under Private deployment model, the Cloud infrastructure is exclusively used by a single organization. In this environment, the organization is in charge of setting up and maintaining the Cloud resources. Accomplishing this requires a significant level of understanding of the organization's business environment and existing resources. However, when done right, there is an added advantage in terms of better control of security, more effective regulatory compliance and improved quality of services.
- *Community Clouds*—Under Community deployment model, the Cloud infrastructure is shared exclusively between organizations from a specific group or community and have common computing concerns. The Cloud framework may be owned, managed and operated by one or more organizations and may be deployed on or off their premises.
- *Hybrid Clouds*—Under Hybrid deployment model, the Cloud infrastructure consists of two or more distinct Clouds (Public, Private or Community). These composite Clouds remain unique entities, but under the Hybrid model, they are bound together by standardized or proprietary technologies that enable data and application portability. By utilizing this model, organizations are able to achieve fault tolerance for their mission-critical processes.

1.2 Benefits of Cloud Computing

Cloud Computing promises numerous benefits, inherent in the characteristics listed in Sect. 1.1.1. According to [4, 9, 10], some of the key benefits offered by Cloud Computing include:

- *Lower cost*—Cloud Computing significantly lowers the cost of entry for firms trying to take advantage of compute-intensive business analytics that were hitherto only available at large corporations. Cloud Computing also represents a huge opportunity to many poor nations that have so far been playing catch-up in the IT revolution.
- *Optimization of capital investment*—Cloud Computing allows companies to optimize their capital investments by reducing the costs of hardware and software purchases. Organizations that have peak requirements can now rent additional hardware on the Cloud instead of having to purchase new equipment. Similarly, instead of purchasing separate software packages for each computer in the organization, Cloud Computing allows IT administrators to host the required software on Cloud, which allows for lower installing and maintenance costs.
- *Rapid scaling*—Cloud Computing allows enterprises to scale their services according to the demands of the customer. Since the computing resources are managed

through software, services can be deployed very quickly as and when new requirements arise.

- *Self service*—Cloud Computing enables consumers to access, configure and deploy
  Cloud services without requiring to interact with any of the service providers. Users
  typically use a service portal provided by the Cloud platform to configure various
  resources and services.
- *Anywhere, anytime access*—Cloud Computing enables true device and location
  independence for its users. Users are no longer bound to a single computer, network or geographic location. Users can access Cloud services using the Internet
  regardless of their location or device type.
- *Multi-tenancy*—Cloud Computing typically allows single instances of software
  applications to serve multiple customers, allowing the service providers to leverage
  on the economies of scale while also reducing maintenance costs.
- *Easier collaboration*—Cloud Computing allows multiple users to easily collaborate, as witnessed by Cloud services like Google Docs and Microsoft Office
  365, which enable users across different geographical locations to collaborate on
  documents, spreadsheets and presentations.
- *Utility service*—Cloud Computing follows a pricing model similar to other utilities, which allows users to pay for only those computing resources that they actually
  used and not for any dedicated resources which may only be used at certain peak
  times.
- *Disaster recovery*—Through use of virtualization [11, 12], Cloud Computing
  delivers faster recovery times and high availability to enterprises, at a fraction of
  the cost of conventional systems. This makes Cloud very attractive for enterprises
  who want to deploy comprehensive disaster recovery plans for their computing
  infrastructure.

1.3 Cloud Computing and Data Centers

In a survey conducted by Cloud.com in the second quarter of 2011, about 61 % of the
organizations surveyed were either in early stages of planning or had already acquired
an approved strategy for implementing Cloud Computing. Furthermore, about 20 %
of the surveyed participants already had Cloud implementations in their organizations
[13]. While the number of organizations leaning towards Cloud related technologies
continues to grow, the general public has already embraced Cloud Computing in
form of services like Office 365 [14], Facebook [15], Flikr [16], Yahoo Applications
[17], Amazon EC2 [18], Youtube [19] and Gmail [20].

Growing adoption of Cloud Computing, by the general public as well as the IT
industry, is driving service providers into deploying new data centers. Data centers
are sites that host hundreds of thousands of servers which concurrently support a
myriad of distinct services and applications [21]. Such facilities, not only let service
providers leverage the economies of scale for bulk deployments, but also allow them

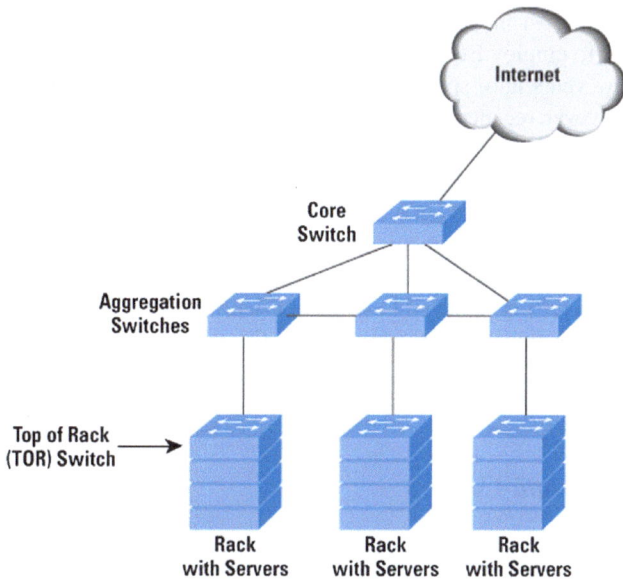

Fig. 1.3 Data center switch network architecture

to dynamically relocate resources among services as workloads change or equipments fail [22, 23].

A data center is generally organized in rows of *'racks'* where each rack contains modular assets such as servers or storage *'bricks'* [24]. These racks are interconnected through *Top-of-Rack* (TOR) switch, which in subsequently connects to the *Aggregation* switch as depicted in Fig. 1.3 from [25].

The Aggregation switch communicates with other Aggregation switches and through them to other servers or storage bricks within the data center. A *Core* switch services various Aggregation switches and provides them with connectivity to the outside world, typically over the Network Layer [26]. It is evident that most of the intra-data center traffic would only traverse the Top-of-Rack and the Aggregation switches [25]. As a result, the overall performance of services and applications within a data center, largely depends on the efficiency of its underlying communication fabric.

As far as building communication fabric for a data center goes, there are essentially two choices. The first option relies on specialized hardware and protocols like Infiniband [27], Myrinet [28] or FibreChannel [29]; the second choice relies on Ethernet [30] based commodity switches that are available off-the-shelf [31]. While the first option is capable of scaling up to thousands of nodes, it is generally more expensive (about $500–$2000 per port [32]) and not natively compatible with TCP/IP applications. On the other hand, the second option is cheap (less than $30 per port [32]), supports a familiar management infrastructure and requires no modification to applications, operating system or system hardware, but scales poorly with increasing number of nodes. Cost and compatibility reasons however, persuade many data center administrators to employ Ethernet as their baseline communication fabric [33].

Until a few years ago, speeds in Ethernet based data centers averaged around 100 Mbps. However, recent revisions to IEEE 802.3 standards have led to the development of Ethernet networks, that have speeds of 1 and 10 Gbps. Today, 1 Gbps Ethernet networks are being widely deployed, and 10 Gbps will be commonly deployed as it becomes affordable. The sudden increase in Ethernet's speeds requires proportional scaling of communication protocols that use it, so that applications that are network intensive can ultimately benefit from the increased bandwidth [34]. Although Internet Protocol (IP) [35] is expected to scale well with the evolving Ethernet, there are some legitimate questions about Transmission Control Protocol (TCP) [36] as noted in [37].

1.3.1 TCP in Data Centers

TCP is a protocol standard that is mature and has survived the test of time. As a standard it has been successfully adapted to several new environments like, long fat networks [38–44], Asynchronous Transfer Mode (ATM) [45] networks [46, 47], as well as wireless and cellular networks [48–58]. However, the unique workloads, speed and scale of modern data centers violate some of the basic assumptions that TCP was originally based upon. For example, in contemporary operating systems such as Linux, the default value of TCP's retransmission timer is set to 200 ms—a reasonable duration for wide area networks where round trip times (RTT) are typically clocked in milli seconds, but two to three orders of magnitude greater than the average round trip time inside data centers [59]. As a result, when TCP is utilized in high-bandwidth, low-latency data center environments, we discover new shortcomings in the protocol. One such shortcoming is referred to as the 'Incast' problem [60].

1.3.1.1 TCP Incast

TCP Incast is a catastrophic collapse in TCP's throughput that occurs in high bandwidth, low latency network environments when multiple senders communicating with a single receiver, collectively send enough data to surpass the buffering abilities of the receiver's Ethernet switch. The problem arises from a subtle interaction between limited Ethernet switch buffer sizes, TCP's loss recovery mechanisms and traffic patterns that are characteristic of data center applications. Small Ethernet buffers get overwhelmed by the large volume of traffic arising concurrently from many servers, which results in packet drops at the switch and one or more TCP timeouts at the servers. These TCP timeouts impose a delay of hundreds of milliseconds at the senders that operate on a network whose round trip time is measured in tens and hundreds of microseconds [61]. As a result of this, the perceived goodput, which can be defined as the data throughput observed at the receiver's application,

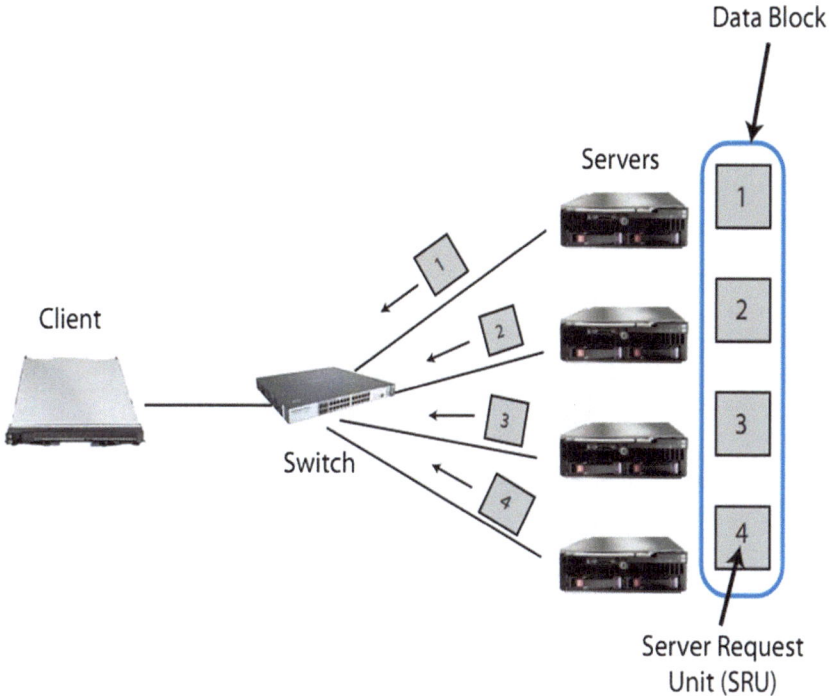

Fig. 1.4 Synchronized reads in cluster storage system

is orders of magnitude lower than the receiver's link capacity. For example, consider a cluster-based storage system discussed in [62]. In a cluster-based storage system, data is typically saved across many storage servers to enhance the reliability and performance of the system. Typically, the networks of such storage systems have low latency (round trip times of 10–100 μs) and high bandwidth (1–10 Gbps). The clients and the servers in such networks are usually separated from each other by one or more switches.

In this environment, the storage severs only store a fragment of the data block since the blocks get striped across a number of servers by the storage system. The fragment of data stored by the servers is denoted as a Server Request Unit (SRU) and is depicted in Fig. 1.4 from [62]. A client requesting a data block from the cluster-based system sends request packets to all the storage servers that contain the SRUs for that particular block; it makes the request for the next block only after receiving all the SRUs for the current requested block. Such requests are referred to as *synchronized reads* in [62].

However, when performing synchronized reads across an increasing number of servers, the requesting client may observe TCP's throughput drop by one or two orders of magnitude below its Ethernet link capacity. Figure 1.5 from [62] illustrates

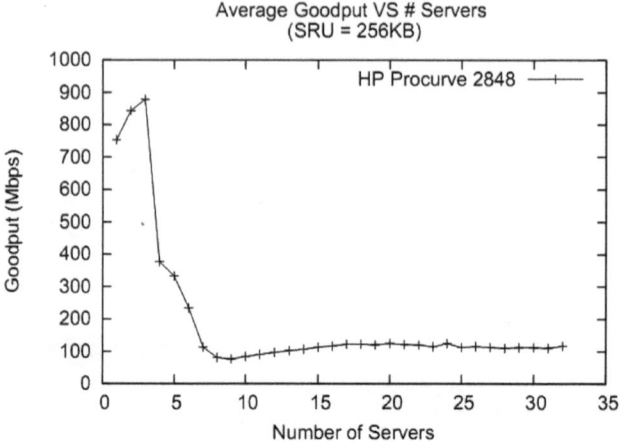

Fig. 1.5 TCP goodput collapse for synchronized reads

TCP's catastrophic drop in performance when operating in a cluster-based storage network environment with HP ProCurve 2848 as the intermediate switch.

Simulation traces reveal that TCP's retransmission timeouts are the primary cause behind Incast. Even when the goodput experienced by the client application degrades, most servers continue to send their SRUs quickly, but some servers begin to experience timeouts from packet losses leading to transmission delays. The servers that finish their transfers receive requests for new SRUs only after the client has completely received its previously requested data block, resulting in underutilized links within the network [63].

Unfortunately, such synchronized read patterns occur frequently in many data center applications and services. For example, in cluster storage when storage nodes respond to requests for data [64–67], in web search when many worker threads respond near simultaneously to query strings [68–71], and in batch processing jobs like MapReduce in which 'mappers' transfer intermediate key-value pairs to appropriate 'reducers' during the 'shuffle' stage [72, 73]. Hence, a feasible solution that addresses the Incast problem is urgently needed.

To the best of our knowledge the problem of Incast has so far never been addressed convincingly. Except for a few attempts in recent literature ([61, 62, 74, 75]), Incast has largely remained unscrutinized. Most existing systems employ solutions that attempt to avoid TCP throughput collapse by capping the number of storage servers involved in a block transfer, by increasing the size of the data blocks, by relying on enhancements to underlying Ethernet technology, or by drastically reducing the value of TCP's minimum retransmission timeout using system extensions to support microsecond clock granularity. Such solutions unfortunately, are specific to a given environment (e.g. a number of servers involved, sizes of data blocks being exchanged, Ethernet support, availability of microsecond timers, etc.), and thus are not robust to any changes in the data center.

Our goal in this study therefore, is to provide practical, backward compatible, transport layer solutions to TCP's Incast problem when operating in high bandwidth, low latency data center network environment.

1.4 Structure of Manuscript

This manuscript is organized as follows: In Chap. 2, we provide an overview of the Transmission Control Protocol, including a brief description of some of its features like reliable delivery, flow control and congestion control. In Chap. 3, we derive a simple analytical model for TCP Incast, followed by its empirical validation. In Chap. 4, we describe techniques to address TCP Incast and evaluate the solutions using simulations. Finally we present our conclusions and directions for future work in Chap. 5.

References

1. S.L. Garfinkel, H. Abelson, *Architects of the Information Society: Thirty-Five Years of the Laboratory for Computer Science at MIT* (The MIT Press, Cambridge, 1999)
2. J. McCarthy, Reminiscences on the history of time sharing. (1983) [Online]. http://www-formal.stanford.edu/jmc/history/timesharing/timesharing.html
3. R. Buyya, C.S. Yeo, S. Venugopal, J. Broberg, I. Brandic, Cloud computing and emerging IT platforms: vision, hype, and reality for delivering computing as the 5th utility. Future Gener. Comput. Syst. **25**, 599–616 (2009) [Online]. http://dx.doi.org/10.1016/j.future.2008.12.001
4. Z. Mahmood, R. Hill, *Cloud Computing for Enterprise Architectures* (Springer Publishing Company (Incorporated), London, 2011)
5. D.W. Cearley, Cloud Computing: Key Initiative Overview, Gartner Report, Gartner, Inc., (2010) [Online]. http://www.gartner.com/it/initiatives/pdf/KeyInitiativeOverview_CloudComputing.pdf
6. J. Rhoton, *Cloud Computing Explained: Implementation Handbook for Enterprises.* (Recursive Press, London, 2009)
7. P. Mell, T. Grance, The NIST Definition of Cloud Computing, Special Publication 800–145, National Institute of Standards and Technology, Technical Report, Sep 2011 [Online]. http://csrc.nist.gov/publications/nistpubs/800-145/SP800-145.pdf
8. W.Y. Chang, H. Abu-Amara, J.F. Sanford, *Transforming Enterprise Cloud Services.* (Springer Publishing Company Incorporated, New York, 2010)
9. S. Marston, Z. Li, S. Bandyopadhyay, J. Zhang, A. Ghalsasi, Cloud computing—the business perspective. Decis. Support Syst. **51**(1), 176–189 (2011) [Online]. http://www.sciencedirect.com/science/article/pii/S0167923610002393
10. M. Miller, *Cloud Computing: Web-Based Applications that Change the Way You Work and Collaborate Online.* (Que, Indianapolis, 2008)
11. C. Takemura, L.S. Crawford, *The Book of Xen: A Practical Guide for the System Administrator.* (No Starch Press, San Francisco, 2009)
12. J. Arrasjid, K. Balachandran, D. Conde, G. Lamb, S. Kaplan, *Deploying the VMware Infrastructure* (The USENIX Association, Berkeley, 2010)
13. Cloud.com, 2011 Cloud Computing Outlook (2011) [Online]. http://www.cloudstack.org/cloud-computing-docs/cloud-computing-survey.pdf

14. K. Murray, *Microsoft Office 365: Connect and Collaborate Virtually Anywhere, Anytime*, 1st edn. (Microsoft Press, Redmond, Washington, 2011)

15. D. Beaver, S. Kumar, H.C. Li, J. Sobel, P. Vajgel, Finding a needle in Haystack: Facebook's photo storage, *Proceedings of the 9th USENIX conference on Operating systems design and implementation, ser. OSDI'10* (USENIX Association, Berkeley, CA, USA, 2010), pp. 1–8 [Online]. http://dl.acm.org/citation.cfm?id=1924943.1924947

16. B.F. Cooper, E. Baldeschwieler, R. Fonseca, J.J. Kistler, P.P.S. Narayan, C. Neerdaels, T. Negrin, R. Ramakrishnan, A. Silberstein, U. Srivastava, R. Stata, Building a cloud for Yahoo!. IEEE Data. Eng. Bull. **32**, 36–43 (2009)

17. B.F. Cooper, R. Ramakrishnan, U. Srivastava, A. Silberstein, P. Bohannon, H.-A. Jacobsen, N. Puz, D. Weaver, R. Yerneni, PNUTS: Yahoo!'s hosted data serving platform. Proc. VLDB Endowment **1**(2), 1277–1288 (2008) [Online]. http://dx.doi.org/10.1145/1454159.1454167

18. D. Robinson, *Amazon Web Services Made Simple: Learn how Amazon EC2, S3, SimpleDB and SQS Web Services Enables you to Reach Business Goals Faster* (Emereo Pty Ltd, London, 2008)

19. J. E. Burgess, Youtube, in *Oxford Bibliographies Online*. (Oxford University Press, Oct 2011), final version following copy-editing by OUP [Online]. http://eprints.qut.edu.au/46719/

20. F.P. Miller, A.F. Vandome, J. McBrewster, *Gmail: Gmail. Webmail, Post Office Protocol, Internet Message Access Protocol, Google, Gmail interface, History of Gmail, Paul Buchheit, Google's hoaxes, Comparison of webmail providers, Gmail Mobile*. (Alpha Press, Orlando, 2009)

21. A. Greenberg, J.R. Hamilton, N. Jain, S. Kandula, C. Kim, P. Lahiri, D.A. Maltz, P. Patel, S. Sengupta, VL2: a scalable and flexible data center network, in *Proceedings of the ACM SIGCOMM 2009 Conference on Data communication, ser. SIGCOMM '09*. (ACM, New York, USA, 2009), pp. 51–62 [Online]. http://doi.acm.org/10.1145/1592568.1592576

22. M. Armbrust, A. Fox, R. Griffith, A.D. Joseph, R.H. Katz, A. Konwinski, G. Lee, D.A. Patterson, A. Rabkin, I. Stoica, M. Zaharia, Above the Clouds: A Berkeley View of Cloud Computing. EECS Department, University of California, Berkeley, Technical Report UCB/EECS-2009-28, Feb 2009 [Online]. http://www.eecs.berkeley.edu/Pubs/TechRpts/2009/EECS-2009-28.html

23. A. Greenberg, J. Hamilton, D.A. Maltz, P. Patel, The cost of a cloud: research problems in data center networks. SIGCOMM Comput. Commun. Rev. **39**(1), 68–73 (2008) [Online]. http://doi.acm.org/10.1145/1496091.1496103

24. K. Kant, Data center evolution: a tutorial on state of the art, issues, and challenges. Comput. Netw. **53**(17), 2939–2965 (2009) Virtualized data centers [Online]. http://www.sciencedirect.com/science/article/pii/S1389128609003090

25. T. Sridhar, Cloud computing: a primer part 1: models and technologies. Internet Protoc. J. **12**(3), 2–19 (2009) [Online]. http://www.cisco.com/web/about/ac123/ac147/archived_issues/ipj_12-3/index.html

26. L.J. Miller, The ISO reference model of open systems interconnection: a first tutorial, in *Proceedings of the ACM '81 conference*, ser. ACM '81. (ACM, New York, USA, 1981), pp. 283–288 [Online]. http://doi.acm.org/10.1145/800175.809901

27. T. Shanley, *Infiniband* (Addison-Wesley Longman Publishing Co., Inc., Boston, 2002)

28. N.J. Boden, D. Cohen, R.E. Felderman, A.E. Kulawik, C.L. Seitz, J.N. Seizovic, W.-K. Su, Myrinet: A gigabit-per-second local area network. IEEE Micro. **15**(1), 29–36 (1995) [Online]. http://dx.doi.org/10.1109/40.342015

29. V. Nagasamy, S. Rajan, P.R. Panda, Fibre channel protocol: formal specification and verification, in *Sixth Annual Silicon Valley Networking Conference*. SysTech Research, (Apr 1995) [Online]. http://www.csl.sri.com/papers/svnc95/

30. IEEE Standard for Information Technology—Telecommunications and Information Exchange Between Systems—Local and Metropolitan Area Networks—Specific Requirements—Part 3: Carrier Sense Multiple Access With Collision Detection (CSMA/CD) Access Method and Physical Layer Specifications, LAN/MAN Standards Committee, New York, NY, USA, (2008) [Online]. http://standards.ieee.org/about/get/802/802.3.html

31. M. Al-Fares, A. Loukissas, A. Vahdat, A scalable, commodity data center network architecture, in *Proceedings of the ACM SIGCOMM 2008 Conference on Data communication*, ser. SIGCOMM '08. (ACM, New York, USA, 2008), pp. 63–74 [Online]. http://doi.acm.org/10. 1145/1402958.1402967

32. U. Hoelzle, L.A. Barroso, *The Datacenter as a Computer: An Introduction to the Design of Warehouse-Scale Machines*, 1st edn. (Morgan and Claypool Publishers, San Rafael, 2009)

33. J. Hamilton, On designing and deploying internet-scale services, in *Proceedings of the 21st Conference on Large Installation System Administration Conference*, ser. LISA'07. (USENIX Association, Berkeley, USA, 2007), pp. 18:1–18:12 [Online]. http://dl.acm.org/citation.cfm? id=1349426.1349444

34. G. Regnier, S. Makineni, R. Illikkal, R. Iyer, D. Minturn, R. Huggahalli, D. Newell, L. Cline, A. Foong, TCP onloading for data center servers. Computer **37**(11), 48–58 (2004) [Online]. http://dx.doi.org/10.1109/MC.2004.223

35. J. Postel, DoD standard Internet Protocol, RFC 760, Internet Engineering Task Force, Jan. 1980, obsoleted by RFC 791, updated by RFC 777 [Online]. http://www.ietf.org/rfc/rfc760.txt

36. Transmission Control Protocol, RFC 793 (Standard), Internet Engineering Task Force, Sep. 1981, updated by RFCs 1122, 3168, 6093, 6528 [Online]. http://www.ietf.org/rfc/rfc793.txt

37. N. Jani, K. Kant, SCTP Performance in Data Center Environments, in *Proceedings of the 2005 International Symposium on Performance Evaluation of Computer and, Telecommunication Systems* (SPECTS'05) (2005)

38. V. Jacobson, R. Braden, D. Borman, TCP Extensions for High Performance. RFC 1323 (Proposed Standard), Internet Engineering Task Force, May 1992 [Online]. http://www.ietf.org/ rfc/rfc1323.txt

39. L.S. Brakmo, S.W. O'Malley, L.L. Peterson, TCP Vegas: new techniques for congestion detection and avoidance, in *Proceedings of the conference on Communications architectures, protocols and applications*, ser. SIGCOMM '94. (ACM, New York, USA, 1994), pp. 24–35 [Online]. http://doi.acm.org/10.1145/190314.190317

40. T. Kelly, Scalable TCP: improving performance in highspeed wide area networks. SIGCOMM Comput. Commun. Rev. **33**(2), 83–91 (2003) [Online]. http://doi.acm.org/10.1145/956981. 956989

41. D.X. Wei, C. Jin, S.H. Low, S. Hegde, FAST TCP: motivation, architecture, algorithms, performance. IEEE/ACM Trans. Netw. **14**(6), 1246–1259 (2006) [Online]. http://dx.doi.org/10. 1109/TNET.2006.886335

42. S. Ha, I. Rhee, L. Xu, CUBIC: a new TCP-friendly high-speed TCP variant. SIGOPS Oper. Syst. Rev. **42**(5), 64–74 (2008) [Online]. http://doi.acm.org/10.1145/1400097.1400105

43. S. Floyd, HighSpeed TCP for Large Congestion Windows. RFC 3649 (Experimental), Internet Engineering Task Force (Dec. 2003) [Online]. http://www.ietf.org/rfc/rfc3649.txt

44. L. Xu, K. Harfoush, I. Rhee, Binary Increase Congestion Control (BIC) for fast long-distance networks, in *IEEE Infocom*. IEEE (2004) [Online]. http://www.ieee-infocom.org/2004/Papers/ 52_4.PDF

45. ATM Forum Inc., ATM User Network Interface (UNI) Specification Version 3.1, 1st edn. (Prentice Hall, Upper Saddle River, New Jersy, 1995)

46. M. Perloff, K. Reiss, Improvements to TCP performance in high-speed ATM networks. Commun. ACM **38**(2), 91–100 (1995) [Online]. http://doi.acm.org/10.1145/204826.204849

47. A. Romanow, S. Floyd, Dynamics of TCP traffic over ATM networks, in *Proceedings of the Conference on Communications Architectures, Protocols and Applications*, ser. SIGCOMM '94. (ACM, New York, USA, 1994), pp. 79–88 [Online]. http://doi.acm.org/10.1145/190314. 190322

48. H. Balakrishnan, S. Seshan, E. Amir, R.H. Katz, Improving TCP/IP performance over wireless networks, in *Proceedings of the 1st Annual International Conference on Mobile Computing and Networking*, ser. MobiCom '95. (ACM, New York, USA, 1995), pp. 2–11 [Online]. http:// doi.acm.org/10.1145/215530.215544

49. S.R. Cho, H. Sirisena, K. Pawlikowski, An end-to-end freeze TCP with timestamps for ad hoc networks, in *ICC 2005, 40th IEEE International Conference on Communications* ed. by

B.G. Lee. IEEE Communications Society (Piscataway, NJ, USA, May 2005), pp. 3576–3582 [Online]. http://dx.doi.org/10.1109/ICC.2005.1495084

50. S.E. Kim, J.A. Copeland, TCP for seamless vertical handoff in hybrid mobile data networks, in *Global Telecommunications Conference, 2003. GLOBECOM '03*. IEEE, vol. 2 (2003), pp. 661–665 [Online]. http://ieeexplore.ieee.org/xpls/abs_all.jsp?arnumber=1258321

51. K. Brown, S. Singh, M-TCP: TCP for mobile cellular networks. SIGCOMM Comput. Commun. Rev. **27**(5), 19–43 (1997) [Online]. http://doi.acm.org/10.1145/269790.269794

52. S.H. Lee, H.G. Ahn, J.S. Lim, S.H. Kwak, S. Kim, Performance analysis of snoop TCP with freezing agent over cdma2000 networks, in *Proceedings of the 7th CDMA International Conference on Mobile Communications*, ser. CIC'02. (Springer-Verlag, Berlin, Heidelberg 2003), pp. 496–505 [Online]. http://dl.acm.org/citation.cfm?id=1766911.1766973

53. E. Hossain, N. Parvez, Enhancing TCP performance in wide-area cellular wireless networks: transport level approaches, in *Wireless Communications Systems and Networks*, ed. by M. Guizani (Plenum Press, New York, USA, 2004), pp. 241–289 [Online]. http://dl.acm.org/citation.cfm?id=1016648.1016658

54. J. Liu, S. Singh, ATCP: TCP for mobile ad hoc networks. IEEE J. Sel. Areas Commun. **19**(7), 1300–1315 (2002) [Online]. http://dx.doi.org/10.1109/49.932698

55. I.F. Akyildiz, G. Morabito, S. Palazzo, TCP-Peach: a new congestion control scheme for satellite IP networks. IEEE/ACM Trans. Netw. **9**(3), 307–321 (2001) [Online]. http://dx.doi.org/10.1109/90.929853

56. C.P. Fu, S.C. Liew, TCP Veno: TCP enhancement for transmission over wireless access networks. IEEE J. Sel. A. Commun. **21**(2), 216–228 (2006) [Online]. http://dx.doi.org/10.1109/JSAC.2002.807336

57. K. Xu, Y. Tian, N. Ansari, TCP-Jersey for wireless IP communications. IEEE J. Sel. A. Commun. **22**(4), 747–756 (2006) [Online]. http://dx.doi.org/10.1109/JSAC.2004.825989

58. E. H.-K. Wu, M.-Z. Chen, JTCP: jitter-based TCP for heterogeneous wireless networks. IEEE J. Sel. A. Commun. **22**(4), 757–766 (2006) [Online]. http://dx.doi.org/10.1109/JSAC.2004.825999

59. Y. Chen, R. Griffith, J. Liu, R.H. Katz, A.D. Joseph, Understanding TCP incast throughput collapse in datacenter networks, in *Proceedings of the 1st ACM Workshop on Research on Enterprise Networking*, ser. WREN '09. (ACM, New York, USA, 2009), pp. 73–82 [Online]. http://doi.acm.org/10.1145/1592681.1592693

60. D. Nagle, D. Serenyi, A. Matthews, The Panasas activescale storage cluster: delivering scalable high bandwidth storage, in *Proceedings of the 2004 ACM/IEEE Conference on Supercomputing*, ser. SC '04. (IEEE Computer Society, Washington, DC, USA, 2004), p. 53 [Online]. http://dx.doi.org/10.1109/SC.2004.57

61. A. Phanishayee, E. Krevat, V. Vasudevan, D.G. Andersen, G.R. Ganger, G.A. Gibson, S. Seshan, Measurement and analysis of TCP throughput collapse in cluster-based storage systems, in *Proceedings of the 6th USENIX Conference on File and Storage Technologies*, ser. FAST'08. (USENIX Association, Berkeley, CA, USA, 2008), pp. 12:1–12:14 [Online]. http://dl.acm.org/citation.cfm?id=1364813.1364825

62. E. Krevat, V. Vasudevan, A. Phanishayee, D.G. Andersen, G.R. Ganger, G.A. Gibson, S. Seshan, On application-level approaches to avoiding TCP throughput collapse in cluster-based storage systems, in *Proceedings of the 2nd International Workshop on Petascale Data Storage: Held in Conjunction with Supercomputing '07*, ser. PDSW '07. (ACM, New York, USA, 2007), pp. 1–4 [Online]. http://doi.acm.org/10.1145/1374596.1374598

63. S. Kulkarni, P. Agrawal, A Probabilistic Approach to Address TCP Incast in Data Center Networks, in *Proceedings of the 2011 31st International Conference on Distributed Computing Systems Workshops*, ser. ICDCSW '11. (IEEE Computer Society, Washington, DC, USA, 2011), pp. 26–33 [Online]. http://dx.doi.org/10.1109/ICDCSW.2011.41

64. S. Ghemawat, H. Gobioff, S.-T. Leung, The Google file system. SIGOPS Oper. Syst. Rev. **37**(5), 29–43 (2003) [Online]. http://doi.acm.org/10.1145/1165389.945450

65. R.Y. Wang, T.E. Anderson, *xFS: A Wide Area Mass Storage File System* (University of California at Berkeley, Berkeley, CA, USA, Technical Report, 1993)

66. G. DeCandia, D. Hastorun, M. Jampani, G. Kakulapati, A. Lakshman, A. Pilchin, S. Siva-subramanian, P. Vosshall, W. Vogels, Dynamo: Amazon's highly available key-value store, in *Proceedings of Twenty-first ACM SIGOPS Symposium on Operating Systems Principles*, ser. SOSP '07. (ACM, New York, USA, 2007), pp. 205–220 [Online]. http://doi.acm.org/10.1145/1294261.1294281

67. F. Schmuck, R. Haskin, GPFS: A shared-disk file system for large computing clusters, in *Proceedings of the 1st USENIX Conference on File and Storage Technologies*, ser. FAST '02. (USENIX Association, Berkeley, CA, USA, 2002) [Online]. http://dl.acm.org/citation.cfm?id=1083323.1083349

68. S. Chakrabarti, M. van den Berg, B. Dom, Focused crawling: a new approach to topic-specific web resource discovery, in *Proceedings of the Eighth International Conference on World Wide Web*, ser. WWW '99. (Elsevier North-Holland, Inc., New York, USA, 1999), pp. 1623–1640 [Online]. http://dl.acm.org/citation.cfm?id=313234.313121

69. J. Cho, H. Garcia-Molina, Parallel crawlers, in *Proceedings of the 11th International Conference on World Wide Web*, ser. WWW '02. (ACM, New York, USA, 2002), pp. 124–135 [Online]. http://doi.acm.org/10.1145/511446.511464

70. J. Luo, Z. Shi, Eliminate redundancy in parallel search: a multi-agent coordination approach, in *Proceedings of the 9th Pacific Rim International Conference on Artificial Intelligence*, ser. PRICAI'06. (Springer-Verlag, Berlin, Heidelberg, 2006), pp. 91–100 [Online]. http://dl.acm.org/citation.cfm?id=1757898.1757912

71. M.D. Dikaiakos, A. Katsifodimos, G. Pallis, Minersoft: Software retrieval in grid and cloud computing infrastructures. ACM Trans. Internet Technol. **12**(1), 2:1–2:34 (2012) [Online]. http://doi.acm.org/10.1145/2220352.2220354

72. J. Dean, S. Ghemawat, MapReduce: simplified data processing on large clusters. Commun. ACM **51**(1), 107–113 (2008) [Online]. http://doi.acm.org/10.1145/1327452.1327492

73. K. Shvachko, H. Kuang, S. Radia, R. Chansler, The Hadoop distributed file system, in *Proceedings of the 2010 IEEE 26th Symposium on Mass Storage Systems and Technologies (MSST)*, ser. MSST '10. (IEEE Computer Society, Washington, DC, USA, 2010), pp. 1–10 [Online]. http://dx.doi.org/10.1109/MSST.2010.5496972

74. V. Vasudevan, A. Phanishayee, H. Shah, E. Krevat, D.G. Andersen, G.R. Ganger, G.A. Gibson, B. Mueller, Safe and effective fine-grained TCP retransmissions for datacenter communication, in *Proceedings of the ACM SIGCOMM 2009 Conference on Data Communication*, ser. SIGCOMM '09. (ACM, New York, USA, 2009), pp. 303–314 [Online]. http://doi.acm.org/10.1145/1592568.1592604

75. M. Alizadeh, A. Greenberg, D.A. Maltz, J. Padhye, P. Patel, B. Prabhakar, S. Sengupta, M. Sridharan, Data center TCP (DCTCP), in *Proceedings of the ACM SIGCOMM 2010 Conference*, ser. SIGCOMM '10. (ACM, New York, USA, 2010), pp. 63–74 [Online]. http://doi.acm.org/10.1145/1851182.1851192

Chapter 2
The Transmission Control Protocol

For over three decades, Transmission Control Protocol (TCP) [1] has been the de-facto transport protocol for a countless number of network applications. So popular is the protocol that according to prior studies, TCP accounts for almost 90 % of the byte count in the Internet [2, 3]. TCP's robustness in a wide variety of network-ing environments is one of the primary reasons for its large scale deployment. The protocol's ability to provide adequate performance to diverse applications has only been possible through continuous study, improvements and modifications, making TCP one of the most active areas of research [4]. In this chapter we provide an overview of the TCP along with brief description of its internal mechanisms like reliable delivery and congestion control that that are key to our understanding of the Incast phenomena.

2.1 Overview

The Internet is a huge network or networks, each implementing the Internet Protocol (IP). IP is the principal communications protocol for transmitting information packets across network boundaries where the source and the destination hosts are identified by fixed length addresses [5]. The design of IP however, assumes that the underlying network infrastructure is inherently unreliable. As a result, IP only provides best effort delivery, meaning, the service it provides is not entirely trustworthy.

User applications however, need reliable, in-order delivery with flow control between two communicating endpoints. One possible approach to follow would be to allow each application to implement its own error detection and recovery mechanism. However, given that the mechanism is needed by many applications, advantages of having a common protocol that provides these functionalities, is immediately appar-ent. Not only would the availability of such a protocol ease the design and implemen-tation of user programs, it would also allow for efficient multiplexing of datagrams

S. Kulkarni and P. Agrawal, *Analysis of TCP Performance in Data Center Networks*,   17
SpringerBriefs in Electrical and Computer Engineering,
DOI: 10.1007/978-1-4614-7861-4_2, © The Author(s) 2014

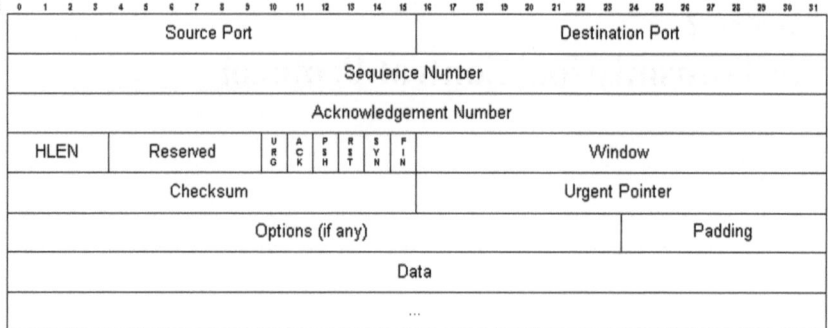

Fig. 2.1 Layout of a TCP Segment

from host to the applications [4]. The TCP was specifically designed to provide such a service.

TCP described in [1] is a connection-oriented, end-to-end reliable protocol designed to fit into the layered hierarchy just above the Internet Protocol. The protocol provides for reliable inter-process communication between pairs of applications in hosts attached to distinct but interconnected communication networks. TCP makes very few assumptions about the reliability of the communication protocols in the layers below itself. TCP only assumes that it has access to a simple and potentially unreliable datagram services from the layers below. This implies that TCP can conceivably perform over a wide spectrum of communication systems ranging from packet-switched networks to hard-wired connections.

Using TCP, applications on networked hosts can create virtual circuits (or connections) to each other, over which they can exchange streams of data. TCP assigns a 32-bit sequence number to every byte of data on its connection. The protocol guarantees reliable, in-order delivery of all data bytes that are sent from the sender to the receiver. TCP also has the ability to distinguish data for multiple connections by concurrent applications.

The sending and receiving TCP endpoints exchange their data in the form of packets that are called segments. A TCP segment consists of a fixed 20-byte header (plus an optional extensions) followed by zero or more data bytes [6]. Figure 2.1 shows the layout of a TCP segment. Table 2.1 lists the purpose of each field in a TCP segment.

The size of the segments exchanged between two endpoints is controlled by the TCP. TCP even decides whether to accumulate data from several writes into one segment or to split data from a single write over several segments. Segment sizes over a given TCP connection is governed by two limits. First, each segment including the segment header must fit into the 65,535 byte IP payload. Second, each network has a Maximum Transfer Unit (MTU), and each TCP segment must fit in this MTU [6].

Table 2.1 TCP segment fields

Field name	Length in bits	Function
Source Port	16	Identifies the local end point at the source
Destination Port	16	Identifies the local end point at the destination
Sequence Number	32	Sequence number of the segment's first data byte in the overall connection byte stream
Acknowledgment Number	32	Sequence number of the next byte expected by the receiver
Header Length	4	Indicates the segment header length in words
URG flag	1	Control flag indicates that the Urgent Pointer field is significant
ACK flag	1	Control flag indicates that the Acknowledgment Number field is significant
PSH flag	1	Control flag requests the receiver to deliver the data to application on arrival
RST flag	1	Control flag used to reset connection
SYN flag	1	Control flag used in establishing connections
FIN flag	1	Control flag to request normal termination of TCP connection in the direction of the segment
Window	16	Used for flow control. Indicates the number of bytes that may be sent starting at the byte acknowledged
Checksum	16	Provides bit error detection for the TCP segment
Urgent Pointer	16	Indicates the position of the first octet of non expedited data in the segment
Options	32*	Zero or more words designed to provide extra facilities not covered by the regular header

*The field is optional

A segment that is too large to fit into the MTU of a network is broken down into multiple fragments by an intermediate router. All resulting fragments get their own IP header and are assembled back into the original segment at the destination.

TCP relies on sliding window protocol to transfer data between two endpoints. When the sender transmits a segment, it also starts a timer called the retransmission timer. On receiving this segment, the destination TCP endpoint sends back a segment bearing an acknowledgment number that indicates the next sequence number it expects to receive from the sender. If the sender's retransmission timer expires before receiving an acknowledgement, the sender would transmit that segment again [6].

Though the operations of TCP sound simple, there are a number of complex situations that the protocol needs to handle. For example, transmitted segments may arrive out of order at the destination. Segments can also get delayed in the network in which case the sender times out and retransmits them. If the retransmitted segments take a different path to the destination, the receiver can end up with multiple copies of the same bytes in the steam. Additionally, if the segment is fragmented, part of the fragmented segments may never arrive at the destination. And last but not the least, a segment may occasionally hit a congested link along its path to the destination.

TCP must be able to handle these situations in an efficient way. A considerable amount of effort has gone into making TCP robust for all network situations. Some of these techniques used by various implementations of TCP will be discussed in the sections below.

2.2 Reliable Data Delivery

In this section, we describe various mechanisms of TCP that are involved in ensuring in-order transfer of stream bytes between source and destination endpoints, as well as, multiplexing of network traffic to different application processes.

Transmission in TCP is made reliable via smart use of sequence numbers and acknowledgments. Conceptually, each byte of data is assigned a sequence number. The sequence number of the segment is also the sequence number of the first byte of data within the segment and is transmitted along with the segment as part of its header. Segment headers also include an acknowledgment number which is the sequence number of the next expected data byte in the opposite direction. When TCP transmits a segment containing data, it puts a copy of the segment on a retransmission queue and starts a timer; when the acknowledgment for the transmitted data is received, it deletes the segment from its retransmission queue. If the acknowledgment is not received before the expiry of the timer, TCP retransmits the segment [1].

In addition to sequence numbers and acknowledgments, TCP's solution for delivering data reliably over an unreliable internet communication system involves the following three mechanisms:

- Establishing connection state at communicating endpoints
- Handling data duplication and reordering
- Handling data loss

The first step in ensuring reliable in-order data delivery between two hosts is the setup of connection state at each endpoints [7] as discussed in the subsection below.

2.2.1 Connection Establishment and Multiplexing

In order to provide reliable data delivery, TCP needs to initialize and maintain certain status information for each connection. The combination of this status information along with sockets, sequence numbers and window sizes, forms a TCP connection or a virtual circuit.

When two network enabled applications wish to communicate, their TCP stacks must first establish a connection by initializing status information at each endpoint. When their communication is complete, the connection is terminated in order to free the resources for other uses [1].

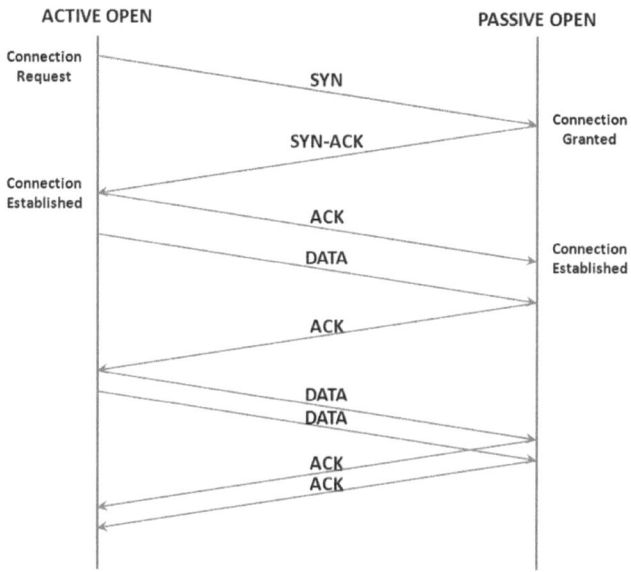

Fig. 2.2 TCP Three-way handshake and initial data exchange

Since connections must be established between processes over the unreliable internet communication system, TCP uses a handshake mechanism with clock-based sequence numbers. The procedure to establish a TCP connection involves exchanging three segments between communicating endpoints, utilizing the synchronize (SYN) control flag in the segment header. This exchange has been termed a three-way hand shake [1] and is depicted in Fig. 2.2. Unlike other connection establishing protocols, three-way handshake does not require communicating endpoints to begin transmissions with same sequence numbers. Furthermore, three-way handshake can be used to establish a TCP connection even in absence of a global clock. The mechanism can also prevent old connection initializations and data packets from causing any confusion. Additionally, the endpoints can exchange parameter and option information such as MSS, during connection establishment [7].

The process which initiates the three-way handshake does so by issuing an *active open* request. Processes can also issue *passive opens* before waiting for matching *active opens* from other networked applications and be informed by TCP when connections have been established. Two applications which issue simultaneous *active* opens to each other will still be correctly connected. This flexibility is critical for TCP to operate in distributed environments where network components can act asynchronously with respect to each other [1].

TCP provides 16-bit port identifiers to distinguish separate data streams that the protocol might handle. Since port identifiers are selected independently by TCP at each communicating endpoint, many endpoints in the network can pick the same identifier for a port. To provide for unique addresses for all communicating processes,

TCP concatenates the IP address identifying the end point with the port identifier that identifies the process, to create a socket which is unique throughout the Internet. A connection in TCP can be fully specified by the pair of sockets at the communicating endpoints.

At each endpoint, the TCP examines the port identifiers in the received segment and places the segment in the receive buffer of the process associated with that port [7]. A range of port identifiers is reserved for well-known user applications such as HTTP [8], FTP [9], SMTP [10], NNTP [11] and SSH [12–17].

2.2.2 Re-ordering and Duplicate Elimination

In this subsection we describe TCP's mechanisms which allow data to be re-ordered at the receiver and duplicate data to be eliminated.

Packet reordering refers to the scenario where relative ordering of some TCP segments belonging to the same connection get altered as they are transported over the network. In other words the receiving order of a stream of segments differs from the sending order.

TCP has the ability to recover from data that is lost, damaged, delivered out of order or duplicated by the network. It achieves this by assigning a sequence number to each transmitted byte and requiring a positive acknowledgment from the receiving endpoint [1]. The receiving endpoint can detect transmission errors by computing a checksum on the received segment and comparing it to the checksum value in the received segment's header. If the checksum test fails, TCP discards the segment. Otherwise, it checks to see if the received sequence number falls within the acceptable range of sequence numbers defined by the receive window, $rwnd$. In TCP, the receive window indicates the allowed number of bytes that the sender may transmit before waiting for new permissions from the receiver.

A data byte whose sequence number does not fall within the sequence number range defined by the receive window is discarded by the TCP. Bytes whose sequence numbers fall within the sequence number range specified by $rwnd$ but are not equal to $rwnd$'s start sequence number are buffered by the TCP. This allows TCP to properly re-order any out of order data. On the other hand, bytes which are received in-order, advance the range boundaries defined by $rwnd$.

Duplicate data in TCP may result from segment duplication by faulty devices, from the finiteness of the sequence space (wrap around), from the presence of segments in the network sent by earlier incarnations of the connection or from retransmissions from the source [7].

In order to limit the possibility of duplicate segments from previous instances of the same connection being erroneously accepted, TCP starts the numbering of data bytes with a "random" value when initiating the connection.

2.2.3 Retransmission of Lost Data

In this subsection we describe TCP's strategy for loss recovery. The strategy employed by TCP mainly relies on positive acknowledgments and timer based retransmissions.

2.2.3.1 Acknowledgments

The receipt of each transmitted byte has to be acknowledged by the receiving end-point. TCP acknowledgment numbers refer to the sequence number of the next byte that the destination expects to receive. This strategy is referred to as "positive acknowledgment" strategy [7]. The acknowledgment mechanism employed by TCP is "cumulative" meaning, an acknowledgment of sequence K indicates that all bytes up to but not including K have been received by the destination. This mechanism allows for simple duplicate detection in presence of retransmissions [1].

If a received segment's sequence number does not match *rwnd*'s current start sequence number, it elicits an acknowledgment for the start sequence number of *rwnd*. Such ACKs, called *duplicate ACKs*, stimulate the sender to retransmit the segment that appears to be missing [7].

It is important to note that an acknowledgment received by the sending endpoint does not guarantee that the data has been delivered to the networked application, but only that the TCP at the receiving endpoint has taken the responsibility to do so.

2.2.3.2 Retransmission Queue

When TCP transmits a segment containing data, it puts a copy of the segment on a queue called retransmission queue and starts a timer that is initialized to a dynamically computed retransmission timeout (RTO) value; when the acknowledgment for that data is received, TCP deletes the segment from this queue. If the acknowledgment is not received before the timer expires, TCP retransmits the segment [1]. Note that segments carrying no data *are not* transmitted reliably, except for segments carrying the SYN or FIN flag.

In addition, a "fast" retransmission of the segment at the head of the retransmission queue can be triggered by the reception of at least three duplicate ACKs before the expiry of the retransmission timer [7]. In both cases the retransmission is followed by congestion control measures that are discussed in Sect. 2.4.

Note that some implementations of TCP, organize the data in retransmit queue in segments, as they were originally transmitted, while others do not keep the segment boundaries. In the first case, when the retransmission timer expires, the segment at the head of the queue is retransmitted. In the second case, a new segment can be created from the data at the head of the retransmission queue. The data in the newly created segment can span over multiple previous segments. This results in more efficient use of the network by decreasing the segment header overhead.

2.3 Flow Control

Flow control is a mechanism whose main purpose is to properly match the transmission rate of the sending end point to that of the receiving end point [18]. TCP uses sliding window mechanism to provide flow control, whereby the receiving end point returns a "window" in each ACK, indicating a range of acceptable data byte sequence numbers beyond the last segment that was successfully received. The window, called receive window or *rwnd*, indicates the allowed number of bytes that the sender may transmit before waiting for new permission from the receiver. Since TCP's *rwnd* field is limited to 16 bits in length, it allows for a maximum possible size of 65,535 bytes.

Figure 2.3 illustrates the concept of the sliding window. In this simple example, the sliding window spans over four bytes of the data stream. The sequence numbers within the sender's window represent the bytes sent but as yet not acknowledged. All sequence numbers to the left of the sliding window are bytes that were transmitted and also acknowledged; sequence numbers to the right of the sliding window are bytes that are yet to be transmitted. As bytes in the window get acknowledged and new bytes get transmitted the window "slides", moving from left to right.

A receiver can adjust the window size each time it sends the acknowledgments to the sender. The maximum transmission rate is ultimately bound by the receiver's ability to accept and process data. If the receiver is incapable of accepting any new

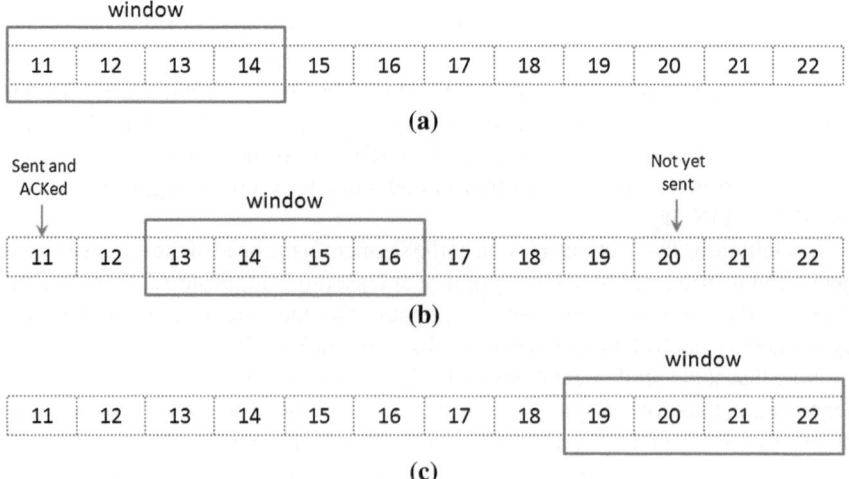

Fig. 2.3 Sliding window mechanism

data, it can announce a "zero receive window" in an ACK, which forces the sender TCP to stall its data transmission.

A sender which receives a zero window advertisement for *rwnd*, regularly probes the receiver for window updates. This is because the underlying IP protocol only provides a best effort service, due to which, an ACK carrying a window update from the receiver can sometimes fail to reach the sender. TCP at the sending endpoint sends the first probe after a retransmit timeout period and sends the subsequent ones at exponentially increasing time periods [19].

TCP at the sending end point also deals with the case where the receiver advertises a window that is smaller than the amount of data already in the network (which corresponded to a previously advertised window value). This case, labeled "shrinking window", causes the sender to wait for the receive window, *rwnd*, to open up beyond the previously sent limit before sending any new data [20].

2.4 Congestion Control

Congestion control in TCP concerns with controlling the entry of segments into the network in order to avoid overwhelming the processing or link capabilities of the intermediate nodes. This section describes TCP's four intertwined algorithms that are implemented as part of the protocol's congestion control strategy: slow start, congestion avoidance, fast retransmit, and fast recovery. The following subsections discuss these algorithms in detail.

2.4.1 Slow Start and Congestion Avoidance

All TCP senders use slow start and congestion avoidance algorithms to control the amount of unacknowledged data being injected into the network. To implement these algorithms, TCP makes use of two variables, namely, congestion window (*cwnd*) and receiver's advertised window (*rwnd*). The congestion window is the sender-side limit on the number of bytes the sender can inject into the network before receiving an acknowledgment, while the receiver's window is a receiver-side limit on the amount of outstanding bytes of data. The minimum of *cwnd* and *rwnd* governs TCP's data transmission.

Another variable, the slow start threshold (*sstrhesh*), is used by the TCP to determine the algorithm to employ—slow start or congestion avoidance—in controlling data transmission.

Starting data transmission with unknown network condition requires TCP to avoid congesting the network with large burst of data. Hence it probes the network slowly and determines the available capacity, using its slow start algorithm. It is either used at the very beginning of data transfer or after repairing loss detected by TCP's retransmission timer. In both these situations, TCP is unaware of the current state of the network causing it to probe the system for available capacity.

Initially, TCP sets *ssthresh* to an arbitrarily high value, but reduces it in response to congestion. Setting *ssthresh* to a high value initially ensures that network conditions, rather than some arbitrary host limit, dictates the sending rate. TCP uses the slow start algorithm when its *cwnd* ≤ *ssthresh* and uses the congestion avoidance algorithm when *cwnd* > *ssthresh*.

During slow start, TCP increments its *cwnd* by at most one maximum segment size (MSS) for each ACK received. Slow start ends when *cwnd* exceeds *ssthresh* or when TCP observes congestion in the network.

In reality, slow start is not very slow when the network is not congested and network response time is good. For example, the first successful transmission and acknowledgment of a TCP segment increases *cwnd* to two segments. After successful transmission and acknowledgment of these two segments, the *cwnd* is doubled to four segments, then eight segments, then sixteen segments and so on, up to the maximum window size (*rwnd*) advertised by the receiver or until TCP observes congestion in the network.

During congestion avoidance, *cwnd* is increased by roughly one MSS per round-trip time. Congestion avoidance continues until congestion is detected. Another common formula that is used by various implementations of TCP in updating *cwnd* during congestion avoidance phase is given in Eq. 2.1.

$$cwnd = cwnd + \frac{(MSS \times MSS)}{cwnd} \tag{2.1}$$

This adjustment to congestion window is executed on every incoming ACK that acknowledges new data during the congestion avoidance phase.

When a TCP sender detects segment loss through expiry of the retransmission timer and the segment in question has not yet been retransmitted, TCP sets the value of its *ssthresh* according to Eq. 2.2. Furthermore, upon a timeout, TCP sets the value of its *cwnd* to one MSS. Therefore, after retransmitting the dropped segment, TCP sender uses slow start algorithm to increase the size of its congestion window (*cwnd*) from one MSS to the new value of *ssthresh*, at which point congestion avoidance again takes over [21].

$$ssthresh = max \left(\frac{segments\ in\ flight}{2}, 2 \times MSS \right) \tag{2.2}$$

2.4.2 Fast Retransmit and Fast Recovery

When the destination receives an out-of-order segment, TCP at the receiving endpoint immediately sends back a duplicate ACK to the sender. Duplicate ACK informs the sender that the destination received a segment that was out-of-order. The acknowledgment number in the duplicate ACK also informs the sender about the byte sequence number that the destination expects. From the sender's perspective, there are a number of problems that can result in duplicate ACKs. For example, duplicate ACKs can be caused by segments getting dropped by the network. In this case, all segments received by the destination after the dropped segment will trigger duplicate ACKs until the loss is repaired. Duplicate ACKs can also be caused by the re-ordering or replication of segments within the network.

TCP's fast retransmit algorithm uses the arrival of three consecutive duplicate ACKs as an indication that the segment has been lost. After receiving three duplicate ACKs, the sender retransmits the lost segment, without waiting for its retransmission timer to go off.

After TCP's fast retransmit algorithm sends the missing segment, the protocol's fast recovery algorithm controls the transmission of new data until the sender receives non-duplicate ACK from the destination. The reason that TCP does not perform slow start at this stage is that in addition to indicating a segment loss, duplicate ACKs also inform the sender that the segments are most likely leaving the network.

TCP implements the fast retransmit and the fast recovery algorithms in the following manner:

- On the first and the second duplicate ACKs received by the sender, TCP sends a segment of previously unsent data provided, the receiver's *rwnd* allows for it. TCP also does not change its *cwnd* to reflect the transmission of these two segments.
- When the third duplicate ACK is received at the sender, TCP sets *ssthresh* to a value given in Eq. 2.2.
- When the third duplicate ACK is received, following the reset of *ssthresh*, TCP sets its *cwnd* to $(ssthresh + 3 \times MSS)$ ensuring that the *cwnd* is artificially inflated by the number of segments that are outstanding in the network.
- For each additional duplicate ACK that the sender receives, TCP increments its *cwnd* by one MSS.
- When finally the sender receives an ACK that acknowledges previously unacknowledged data, TCP sets *cwnd* to *ssthresh*. This sequence is also known as "deflating" of the congestion window (*cwnd*).

A summary of TCP's congestion control mechanisms is depicted in Fig. 2.4. An illustration of how TCP's congestion window evolves due to the protocol's aforementioned congestion control algorithms, is shown in Fig. 2.5.

In Fig. 2.5, TCP begins by setting its slow start threshold, *ssthresh*, to an arbitrarily high value. It then starts its data transfer using the slow start algorithm to determine the available capacity in the network. During this phase, TCP's congestion window *cwnd*, grows exponentially. In the example above, slow start phase ends when TCP experiences a timeout. Following the timeout, TCP sets its *ssthresh*, to half the number of segments that were in flight before the timeout. The protocol also sets the size of its *cwnd* to one. Since *cwnd* is now less than *ssthresh*, TCP resumes its data transfer with slow start. Like before, *cwnd* grows exponentially as long as $cwnd \leq ssthresh$. When $cwnd > ssthresh$, TCP's slow start phase ends. TCP then continues with its data transfer using the congestion avoidance algorithm. During this phase, *cwnd*

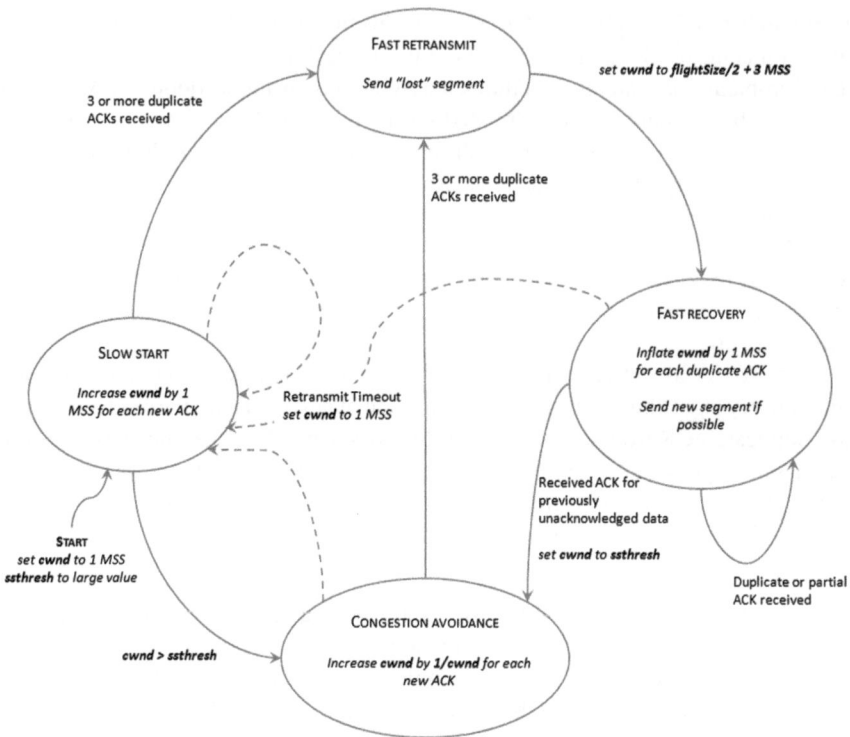

Fig. 2.4 Summary of TCP's congestion control mechanisms

grows linearly until TCP receives 3 duplicate ACKs. On receiving 3 duplicate ACKs, TCP ends its congestion avoidance phase and invokes fast retransmit and fast recovery algorithms. This congestion avoidance-fast retransmit-fast recovery cycle continues until TCP experiences another timeout. Following a timeout, TCP resumes its data transfer with slow start algorithm as before. The resulting evolution pattern for TCP's congestion window *cwnd*, is often referred to as "TCP's sawtooth behavior".

2.5 Summary

In this chapter, we presented details on mechanisms that are responsible for TCP's reliable data transfer, flow control and congestion control. Our goal in this chapter is to not only provide the necessary background for the following chapters, but to also help readers working with TCP to gain a better understanding of the protocol.

We note that TCP is a highly dynamic protocol, especially when the details of its implementations are considered. Many developers independently add non-standard modifications and enhancements to standard implementations of TCP. Moreover,

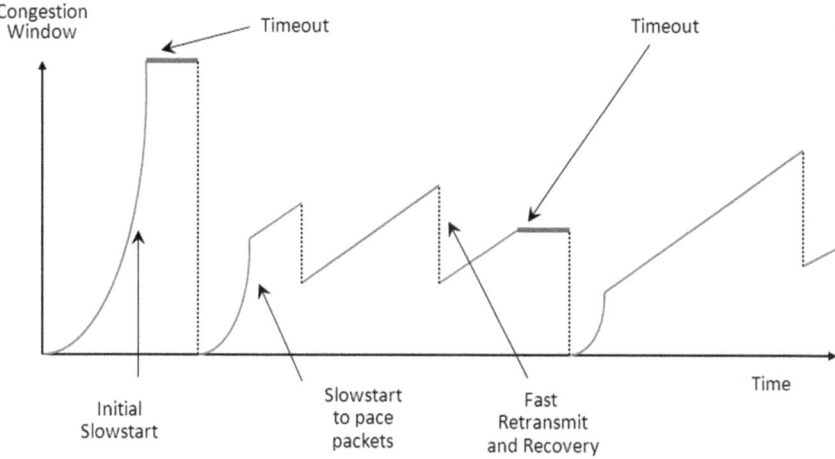

Fig. 2.5 Evolution of TCP's congestion window

due to the complexity of the protocol and some ambiguity in its specification, many developers allow themselves the freedom to deviate from the standard behavior to provide simplicity or inter-operability with other implementations of TCP. Therefore the information contained in this chapter may not apply to every implementation of TCP.

References

1. J. Postel, Transmission control protocol, RFC 793 (Standard), Internet engineering task force, updated by RFCs 1122, 3168, 6093, 6528. (1981). [Online]. Available: http://www.ietf.org/rfc/rfc793.txt
2. K. Thompson, G.J. Miller, R. Wilder, Wide-area Internet traffic patterns and characteristics. Netw. Mag. Glob. Internetworking. **11**(6), pp. 10–23, (1997). [Online]. Available: http://dx.doi.org/10.1109/65.642356
3. S. McCreary and k. claffy, Trends in wide area IP traffic patterns-a view from Ames Internet Exchange, in ITC Specialist Seminar, Monterey, Sept 2000.
4. W. Noureddine, F. Tobagi, The transmission control protocol. [Online]. Available: http://citeseer.ist.psu.edu/noureddine02transmission.html
5. J. Postel, DoD standard internet protocol, RFC 760, Internet engineering task force, Obsoleted by RFC 791, updated by RFC 777. Jan (1980). [Online]. Available: http://www.ietf.org/rfc/rfc760.txt
6. A. Tanenbaum, Computer networks, 4th edn. Prentice Hall Professional Technical Reference, 2002
7. W. K. Noureddine, Improving the performance of tcp applications using network-assisted mechanisms, Ph.D. dissertation, Stanford University, Stanford, 2002, aAI3048586
8. T. Berners-Lee, R. Fielding, H. Frystyk, Hypertext transfer protocol-HTTP/1.0, RFC 1945 (Informational), Internet engineering task force, May 1996. [Online]. Available: http://www.ietf.org/rfc/rfc1945.txt

9. J. Postel, J. Reynolds, File transfer protocol, RFC 959 (Standard), Internet engineering task force, updated by RFCs 2228, 2640, 2773, 3659, 5797. Oct 1985. [Online]. Available: http://www.ietf.org/rfc/rfc959.txt

10. J. Postel, Simple mail transfer protocol, RFC 821 (Standard), Internet engineering task force, obsoleted by RFC 2821. Aug 1982. [Online]. Available: http://www.ietf.org/rfc/rfc821.txt

11. C. Feather, Network news transfer protocol (NNTP), RFC 3977 (Proposed standard), Internet engineering task force, updated by RFC 6048. Oct 2006. [Online]. Available: http://www.ietf.org/rfc/rfc3977.txt

12. T. Ylonen, C. Lonvick, The secure shell (SSH) protocol architecture, RFC 4251 (Proposed standard), Internet engineering task force, Jan 2006. [Online]. Available: http://www.ietf.org/rfc/rfc4251.txt

13. T. Ylonen, C. Lonvick, The secure shell (SSH) authentication protocol, RFC 4252 (Proposed standard), Internet engineering task force, Jan 2006. [Online]. Available: http://www.ietf.org/rfc/rfc4252.txt

14. T. Ylonen, C. Lonvick, The secure shell (SSH) transport layer protocol, RFC 4253 (Proposed standard), Internet engineering task force, Jan 2006. [Online]. Available: http://www.ietf.org/rfc/rfc4253.txt

15. T. Ylonen, C. Lonvick, The secure shell (SSH) connection protocol, RFC 4254 (Proposed standard), Internet engineering task force, Jan 2006. [Online]. Available: http://www.ietf.org/rfc/rfc4254.txt

16. J. Schlyter, W. Griffin, Using DNS to securely publish secure shell (SSH) key fingerprints, RFC 4255 (Proposed standard), Internet engineering task force, Jan 2006. [Online]. Available: http://www.ietf.org/rfc/rfc4255.txt

17. F. Cusack, M. Forssen, Generic message exchange authentication for the secure shell protocol (SSH), RFC 4256 (Proposed standard), Internet engineering task force, Jan 2006. [Online]. Available: http://www.ietf.org/rfc/rfc4256.txt

18. J. Kristoff, The transmission control protocol. [Online]. Available: http://condor.depaul.edu/jkristof/technotes/tcp.html

19. R. Braden, Requirements for internet hosts-communication layers, RFC 1122 (Standard), Internet engineering task force, updated by RFCs 1349, 4379, 5884, 6093, 6298, 6633. Oct 1989. [Online]. Available: http://www.ietf.org/rfc/rfc1122.txt

20. W.R. Stevens, TCP/IP illustrated (vol. 1): the protocols. (Boston, Addison-Wesley Longman Publishing Co., Inc., 1993)

21. M. Allman, V. Paxson, E. Blanton, TCP congestion control, RFC 5681 (Draft standard), Internet engineering task force, Sept 2009. [Online]. Available: http://www.ietf.org/rfc/rfc5681.txt

Chapter 3
Modeling Incast and its Empirical Validation

Transmission Control Protocol is the transport layer workhorse for several application layer protocols like HTTP [1], FTP [2], SMTP [3], NNTP [4] and SSH [5–10]. As a result, TCP carries a significant amount of today's Internet traffic [11]. Studies have shown that traffic from TCP and UDP [12] make for more than 96 % of the packets in the Internet. TCP alone accounts for almost 82 % of packets and about 91 % of the byte count on the Web [13].

TCP also accounts for the bulk of traffic in data centers. TCP is at the core of several data center applications like distributed filesystems [14, 15], cluster comput- ing [16, 17], parallel databases [18] as well as disaster recovery [19, 20]. However, recent works have shown that under certain many-to-one traffic patterns, data center networks experience Incast: a drastic collapse in throughput due to TCP timeouts triggered by severe packet losses at Ethernet [21] switches [22–24].

In typical Incast communication pattern, a receiver issues synchronized data requests to multiple senders. The senders, upon receiving the request, concurrently transmit a large amount of data to the receiver. The data from all senders traverse a bottleneck link in a many-to-one fashion. As the number of concurrent senders increase, the perceived application-level throughput at the receiver collapses. The application at the receiver sees throughput that is orders of magnitude lower than its link capacity [25]. TCP throughput collapse was first observed in early parallel network storage projects such as NASD [26]. It was later documented as part of a larger paper by Nagle et al in [27]. Today, the same Incast communication pattern can be found in many popular data center applications such as cluster based storage systems [27–29], data analytics [30–32], Big Data [33], MapReduce [34] as well as Hadoop [35]. Hence a thorough solution that addresses the Incast pathology is urgently needed.

To substantially solve TCP Incast at low cost, we first need to understand the reasons behind its throughput collapse. Traditionally, simulation and implementa- tion/measurement have been tools of choice for examining the performance of various aspects of TCP. In this chapter we develop a simple analytic characterization of the steady state throughput of multiple TCP flows, as a function of loss rate and round

S. Kulkarni and P. Agrawal, *Analysis of TCP Performance in Data Center Networks*, 31
SpringerBriefs in Electrical and Computer Engineering,
DOI: 10.1007/978-1-4614-7861-4_3, © The Author(s) 2014

trip time under many-to-one Incast communication pattern. Although many earlier works have already modeled TCP [36–41], our modeling is different in two aspects:

1. The application in our model exhibits Incast communication pattern whereas existing models usually assume that the application layer has infinite amount of data to send.
2. Our model describes the overall throughput of the bottleneck link which contains multiple flows, while existing TCP models usually focus on the throughput of a single flow.

In our TCP Incast model, we summarize that the throughput collapse in many-to-one communication pattern is mainly caused by two kinds of timeouts.

- Anterior Block Transfer Timeout (ABTT): Anterior Block Transfer Timeouts happen when a large number of senders get involved in a many-to-one synchronized data transfer. During the transfer of a block, some senders finish transmitting their blocks early due to TCP's unfairness at small timescales. Such completed flows wait for other senders to finish transmitting their blocks, without consuming any of the available bandwidth. Meanwhile the remaining flows finish transmitting their blocks using additional bandwidth vacated by the completed flows. This results in larger transmission window for some flows by the end of the block transfer. At the beginning of the next block transfer, all senders inject their whole windows into the network overwhelming the small buffers at the intermediate Ethernet switch. This results in a lot of dropped packets and if any flow loses all the packets in its window, then it will enter a timeout period.
- Intermediate Block Transfer Timeout (IBTT): Unlike Anterior Block Transfer Timeouts, Intermediate Block Transfer Timeouts are not limited to the start of a block transfer. IBTTs are caused when a participating sender fails to receive enough duplicate ACKs to trigger Fast Recovery following the loss of transmitted packets during a block transfer. The sender waits for a period of time defined by TCP's timeout before retransmitting its unacknowledged packets. Following a timeout, the congestion window is reduced to one, and only one packet is resent in the first round after the timeout. However, because of the synchronized nature of the Incast traffic, the receiver cannot issue its next request until all the senders have finished transmitting their current blocks.

Investigating the causes behind the aforementioned category of timeouts is beneficial in developing effective solutions that are capable of avoiding the ill effects of TCP Incast.

3.1 Modeling Incast

More than a decade after its publication in [36], the steady state throughput equation of TCP by Padhye et al. remains the most widely used method for calculating the throughput that a TCP sender achieves under certain environmental conditions. While there now is a wealth of other models available (e.g. [37–41]), many of which are better in some aspect, none of them seem to strike the same balance between precision and ease of use that makes equation from [36] the useful tool that it is.

In an effort to enable practical calculation of the throughput in Incast communication pattern, we extend the equation from [36] to multiple synchronized TCP flows across a single bottleneck link. We do this by following the basic approach in [36], but considering a number of synchronized flows using an identical path at the same time instead of a single flow.

3.1.1 Model Using Loss Measure of Cumulative Flow

In order to derive an equation for the throughput of Incast traffic, we extend the model presented in [36] to multiple synchronized flows. We assume that the reader is familiar with [36] and therefore will only repeat the preliminary assumptions where needed and shortly repeat necessary definitions.

Consider n parallel TCP flows $f_1, \ldots, f_n$ sharing the same bottleneck link inside a data center network. Like in [36], we too model the congestion avoidance phase of these n flows in terms of "rounds", assuming furthermore that the flows are synchronized in rounds (i.e. in a round, all flows send packets in their current congestion window before the next round starts for all of them). For each flow f, the round starts with the back-to-back transmission of W_f packets, where W_f is the size of the flow's current congestion window. Once all packets falling within the congestion window of all n flows have been sent in this back-to-back manner, no other packets are sent until each flow f, receives an ACK for one of its W_f packets already sent. The first ACK reception by all senders marks the end of the current round and the beginning of the next round. In this model, the duration of a round is equal to the round trip time and is assumed to be independent of the window size. Note that another assumption here is that the time needed to send all the packets in a window is smaller than the round trip time.

At the beginning of the next round, a group of W'_f new packets will be sent by each flow f, where W'_f is the new size of the flow's TCP congestion window. Assume that the receiver acknowledges every packet received with an ACK. Many TCP receiver implementations can be configured to send one ACK for every packet received. If W_f packets are sent by a flow f, in the first round and all are received and acknowledged correctly, then the flow will receive W_f acknowledgments. Since each acknowledgment increases the flow's congestion window size by $\left(\frac{1}{W_f}\right)$, the congestion window size for the flow f, at the beginning of the next round is $W'_f = W_f + 1$. That is, during congestion avoidance and in the absence of loss, the congestion window size of each flow increases linearly in time, with the slope of one packet per round trip time.

Loss of packets in TCP can be detected in one of two ways, either by the reception of three "duplicate ACKs" by the sender or via timeouts. We denote the former event as a "TD" for triple-duplicate ACK loss indication and "TO" for timeout loss indication. When the loss indicating event is a TD, the composite flow f reduces its congestion window W_f, by a factor of two. On the other hand if the loss indication is of type TO, the composite flow f waits for a period of time denoted by T_0 and then reduces the size of its congestion window W_f to one before retransmitting its unacknowledged segments.

As in [36], we too assume that a packet is lost in a round independently of any packets lost in *other* rounds. On the other hand we assume that packet losses are correlated among the back-to-back transmissions within a round, i.e., if a packet is lost, all remaining packets transmitted until the end of that round, irrespective of which flow they belong to, are also lost. This bursty loss behavior which has been shown to arise from the drop-tail queuing discipline in [42, 43], perfectly matches the queue management policy of the Ethernet switches used in data center networks.

3.1.1.1 Cumulative Flow

Now, consider F to be the cumulative flow of n parallel, synchronized TCP flows $f_1, \ldots, f_n$, sharing the same bottleneck link in a data center network. Let W be the cumulative window size of all n composite flows. Because the composite flows are all synchronized, W is essentially the sum of all n congestion windows. As with the single sender, each round starts with the back-to-back transmission of a total of W packets belonging to n flows. If all W packets are sent, received and acknowledged correctly, then the participating n flows will together receive W acknowledgments. Since each acknowledgment increases the individual flow f's congestion window size by $\left(\frac{1}{W_f}\right)$, the cumulative congestion window size at the beginning of the next round is $W' = \sum_{i=1}^{n}(W_{f_i}+1)$, which implies, $W' = W + n$, as there are n parallel flows involved. This means that, when all n composite flows are in congestion avoidance phase and none of them experience a loss, the cumulative window size of all n flows increases linearly in time, with the slope of n packets per round trip time.

Note that we have assumed the packets lost in the same round to be correlated (i.e., if a packet is lost, all remaining packets transmitted until the end of that round, irrespective of which flow they belong to, are also lost). Hence, more than one among n composite flows could potentially experience a loss event in the same round. But TCP flows that experience a loss, reduce their congestion window only once per round trip time. Since the flows are all synchronized in terms of rounds, the resulting cumulative window W, is also modified only once per round trip time. Hence, in the event of correlated losses, recognizing a packet loss in a composite flow f, serves as a loss event indicator for the cumulative flow F. We define TD-period (*TDP*) for the cumulative flow F, as the period between two consecutive loss event indicators. For the ith TD-period, TDP_i, we define A_i be the duration of the period.

A sample path of the evolution of the cumulative window W is shown in Fig. 3.1. Between two TD loss indications, the composite flows are all in congestion avoidance and the cumulative window increases by n packets per round, as discussed above.

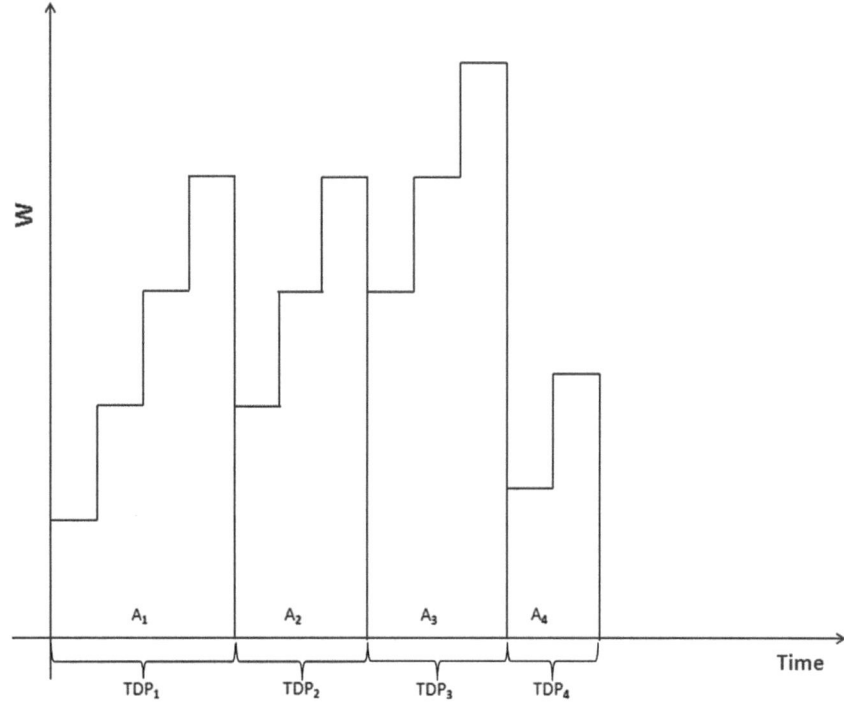

Fig. 3.1 Evolution of cumulative window W over time when loss indications are TDs

Immediately after a loss indication occurs, any composite flow f experiencing a loss, reduces its congestion window size W_f by a factor of two. This implies that a loss experiencing flow f, will also reduce the cumulative window W, by $\left(\frac{W_f}{2}\right)$ packets.

In the following subsections, we model the cumulative flow's behavior in the presence of packet losses. We develop a stochastic model of the cumulative flow corresponding to its operating regimes: when loss indications are exclusively TD and when loss indications are both TD and TO. During the process, we ignore certain aspects of TCP's behavior (e.g. slow start) but believe that we have still managed to captured the essential elements of the protocol, as indicated by the generally good fits between model predictions and simulations, as discussed in Sect. 3.2.

3.1.1.2 Triple Duplicate Loss Indications

In this subsection we assume that loss indications are exclusively of type "triple-duplicate" ACK (TD), and that the composite flow f's window size is not limited by the receiver's advertised flow control window.

For any given time $t \geq 0$, we define N_t as the total number of packets transmitted by the cumulative flow F, in the interval $[0, t]$. Let $B_t = \left(\frac{N_t}{t}\right)$ be the cumulative throughput of all n composite flows in that interval. We can then define the long term steady-state throughput of all n flows as,

$$B = \lim_{t \to \infty} B_t$$
$$= \lim_{t \to \infty} \left(\frac{N_t}{t}\right) \tag{3.1}$$

Note that B_t is the number of packets sent per unit of time regardless of their eventual fate (i.e., whether they are received or not). Thus B_t represents the throughput of the cumulative flow F, at the shared link.

For our new extended equation, we define p_c as the probability of a loss event of the cumulative flow F. It is only counted as a loss event when one or more composite flows f, experiences a loss in a round.

As discussed in the previous subsection, in a loss event of the cumulative flow F, more than one composite flow f, could experience packet loss. In order to estimate the number of composite flows that experience packet loss for each cumulative loss event, we will also use information about real loss probability in our extended equation. With p_r, we denote the probability that a packet (belonging to any composite flow) is lost, given that either it is the flow's first packet in its round or the flow's preceding packet in its round is not lost.

In this subsection, we are interested in establishing a relationship $B(n, p_c, p_r)$ between the throughput of the cumulative flow F and n the number of parallel synchronized flows involved, p_c the loss probability of the cumulative flow as well as p_r the loss probability in any composite flow f.

For a period TDP_i, let Y_i be the number of packets sent in that period and A_i be the duration of that period. From [36], it can be shown that,

$$B = \frac{E[Y]}{E[A]} \tag{3.2}$$

where $E[Y]$ and $E[A]$ are the expected values of Y and A respectively. Hence, to derive B, the longterm steady-state throughput of the cumulative flow, we must next derive the expressions for the mean of Y and mean of A. To achieve this, we need to take a closer look at how the evolution of window size W_f of each composite flow, the time between two loss events of a flow A_f and the duration of a TD-period of each individual flow f, influence the development of the cumulative window size W.

As in [36], we define r_{ij} to be the duration (round trip time) of the jth round of TDP_i and X_i to be the number of rounds in TDP_i. Then, the duration of TDP_i can be computed as $A_i = \sum_{j=1}^{X_i} r_{ij}$. We consider the round trip times r_{ij} to be random variables, that are assumed to be independent of the size of the cumulative window W, and thus independent of the round number, j. It follows that

$$E[A] = E[X]E[r] \tag{3.3}$$

Henceforth, we denote by $RTT = E[r]$, the average value of round trip time.

Since we are now dealing with the cumulative flow, in a single loss event in F, more than one composite flow can experience loss. Let j_i be the number of flows, belonging to the cumulative flow, that experience loss at the end of the ith TD-period. Assuming that loss is identically distributed over all flows, the probability that a composite flow experiences a loss in the ith TD-period is $\left(\frac{j_i}{n}\right)$.

The probability that the time between two loss events of a composite flow A_f, is k TD-periods ($k = 1, 2,\ldots$) is equal to the probability that the flow did not lose a packet in $k - 1$ consecutive TD-periods and in the kth period it loses a packet:

$$P[loss\ in\ the\ kth\ TDP] = \frac{j_i}{n} \prod_{l=1}^{k-1} \left(1 - \frac{j_{(i-l)}}{n}\right) \tag{3.4}$$

If j is the mean number of composite flows experiencing a loss in a round, we have:

$$P[A_f = kE[A]] = \frac{j}{n}\left(1 - \frac{j}{n}\right)^{k-1} \tag{3.5}$$

The mean value of A_f, the time between two loss events for a composite flow, is:

$$E[A_f] = \sum_{k=1}^{\infty} \left(\frac{j}{n}\left(1 - \frac{j}{n}\right)^{k-1} kE[A]\right) \\ = \left(\frac{nE[A]}{j}\right) \tag{3.6}$$

From 3.3 and 3.6 we can express the average number of rounds between two loss events of a flow as:

$$E[X_f] = \frac{nE[X]}{j} \tag{3.7}$$

For deriving Y, we will examine the evolution of the cumulative window W, as shown in Fig. 3.2. In each round, the composite window W, is incremented by n. α_i denotes the sequence number of the first packet lost in TDP_i (for simplicity, we assume the sequence numbers to begin at 1 for every TD-period). After receiving a triple duplicate acknowledgment for one of the composite flows, the cumulative flow recognizes that a packet has been lost (receiving the ACK for packet γ_i). We consider that a TD period ends when the cumulative flow recognizes a loss event. This usually happens in the round following the actual loss; we call this round the "loss round". The total number of packets sent in X_i rounds in TDP_i is $Y_i = \gamma_i$, hence

$$E[Y] = E[\gamma] \tag{3.8}$$

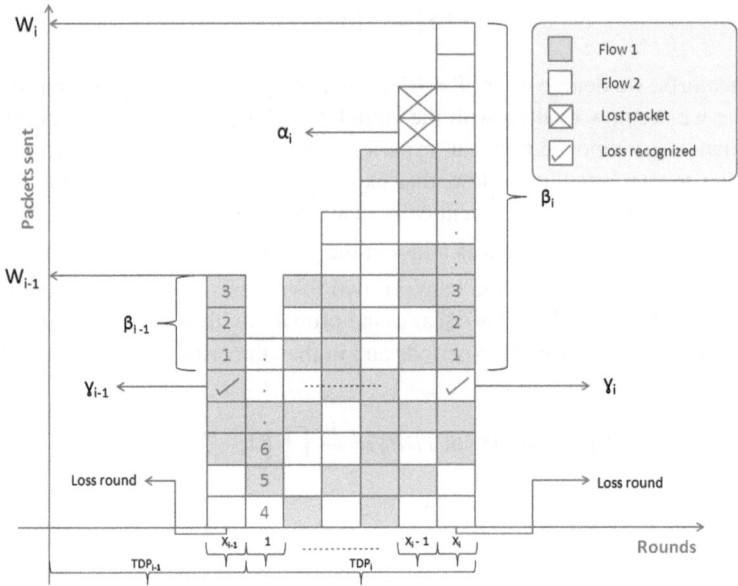

Fig. 3.2 Packets sent during a TD period

The probability that $\gamma_i = k$ is equal to the probability that $k-1$ packets are not loss indications and the ACK for the kth packet triggers the fast retransmission in one of the the composite flows of the cumulative flow F:

$$P[\gamma_i = k] = (1 - p_c)^{k-1} p_c, k = 1, 2, \ldots \qquad (3.9)$$

And the mean value of γ is:

$$E[\gamma] = \sum_{k=1}^{\infty} (1 - p_c)^{k-1} p_c k \qquad (3.10)$$

$$= \left(\frac{1}{p_c}\right)$$

For the ith TD-period let flows x_e, $e = 1, \ldots, j_i$ (subset of n composite flows) be the j_i flows experiencing loss at the end of the period. The same x_e flows do not experience loss in every TD-period. Instead, the TD-periods in which these x_e flows experience loss are a subset ($\{i_s\}$, $s = 1, 2, \ldots$) of TD-periods of the cumulative flow F. For example, in Fig. 3.2, only flow f_2 experiences loss in TDP_i. Its next loss could perhaps happen in the period TDP_{i+2}.

If $W_{f_{x_{e_{i_s}}}}$ are the congestion windows of the flows x_e at the end of the (i_s)th period, and $X_{f_{x_{e_{i_s}}}}$ is the number of rounds from the end of $TDP_{i_{s-1}}$ till the end of TDP_{i_s}, during these $X_{f_{xe_{is}}}$ rounds, the congestion window of flows x_e increase by $X_{f_{xe_{is}}}$ packets. Hence, we have:

$$W_{f_{xe_{is}}} = \frac{W_{f_{xe_{is-1}}}}{2} + X_{f_{xe_{is}}} \tag{3.11}$$

Assuming that $X_{f_{xe_{is}}}$ and $W_{f_{xe_{is}}}$ are mutually independent sequences of independent and identically distributed (i.i.d.) random variables, from [41] we have:

$$E[W_f] = 2E[X_f] \tag{3.12}$$

Assuming that at the end of each TD-period the window sizes of the j flows experiencing loss are $E[W_f]$, and the window sizes of the j flows experiencing loss in the previous loss events are $\left(\frac{E[W_f]}{2} + E[X]\right)$, $\left(\frac{E[W_f]}{2} + 2E[X]\right)$, $\left(\frac{E[W_f]}{2} + 3E[X]\right)$ and so on, the mean window size of the cumulative flow is:

$$E[W] = jE[W_f] + \sum_{k=1}^{\frac{n}{j}-1} j\left(\frac{E[W_f]}{2} + kE[X]\right) \tag{3.13}$$

From 3.7, 3.12 and 3.13, we have:

$$E[W] = \frac{nE[X]}{2} + \frac{3n^2E[X]}{2j} \tag{3.14}$$

The number of packets sent in a TD-period by the cumulative flow F, is the number of packets sent between its two loss events. For the ith TD-period this includes packets sent in the last round of the $(i-1)$th TD-period, starting from the $\gamma_{(i-1)}$th packet till the end of the window (β_{i-1} packets) and the packets sent in the next X_i rounds till the γ_ith packet. If flows x_e, $e = 1,\ldots,j_i$ experience loss in the $(i-1)$th TD-period and $W_{f_{xe_{i-1}}}$ are their respective congestion window sizes at the end of the $(i-1)$th TD-period, the window size of the cumulative flow at the beginning of the ith TD-period is $W_i = \left(W_{i-1} - \sum_{e=1}^{j_{i-1}} \frac{W_{f_{xe_{i-1}}}}{2} + (n - j_{i-1})\right)$, where, j_i flows reduce their congestion windows by factor of two while the remaining $(n - j_{i-1})$ flows increase their window size by one segment. Additionally, the window size W of the cumulative flow F is increased by n every round of the ith TD-period. So the number of packets sent in a TD-period can be expressed as:

$$Y_i = \beta_{i-1} + \sum_{k=0}^{X_i-1} \left(W_{i-1} - \sum_{e=1}^{j_{i-1}} \frac{W_{f_{xe_{i-1}}}}{2} + (n - j_{i-1}) + nk\right) - \beta_i \tag{3.15}$$

where β_i is the number of packets sent in the loss round after the loss event is recognized. Assuming that loss events in the cumulative flow are uniformly distributed over the size of the cumulative window W in a loss round, we have:

$$E[\beta] = \frac{E[W]}{2} \tag{3.16}$$

From 3.15 and 3.16, we can show that:

$$E[Y] = \left(E[W] - \frac{jE[W_f]}{2} + (n - j)\right)E[X] + \frac{nE[X]^2}{2} - \frac{nE[X]}{2} \tag{3.17}$$

and including 3.7, 3.8, 3.10 and 3.12:

$$\frac{1}{p_c} = \frac{3n^2E[X]^2}{2j} + \frac{nE[X]}{2} - jE[X] \tag{3.18}$$

Solving the equation in 3.18 for E[X], we get

$$E[X] = \frac{2j^2p_c - np_cj + \sqrt{n^2p_c^2j^2 - 4np_c^2j^3 + 4j^4p_c^2 + 24n^2p_cj}}{6n^2p_c} \tag{3.19}$$

Including 3.14, we have:

$$E[W] = \frac{2j^2p_c - np_cj + \sqrt{n^2p_c^2j^2 - 4np_c^2j^3 + 4j^4p_c^2 + 24n^2p_cj}}{4p_cj}$$
$$+ \frac{2j^2p_c - np_cj + \sqrt{n^2p_c^2j^2 - 4np_c^2j^3 + 4j^4p_c^2 + 24n^2p_cj}}{12np_c} \tag{3.20}$$

From 3.2, 3.3, 3.10 and 3.19 we can express B, the longterm steady state throughput of all n synchronized flows as:

$$B = \frac{1}{RTT} \times \frac{6n^2}{2j^2p_c - np_cj + \sqrt{n^2p_c^2j^2 - 4np_c^2j^3 + 4j^4p_c^2 + 24n^2p_cj}} \tag{3.21}$$

Equation 3.21 gives us an expression to compute the throughput of Incast traffic when all the composite flows are in congestion avoidance phase and receive only loss indicating events that are of type TD. In this equation, we can approximate j, the mean number of flows experiencing a loss in a round, with the expression $\left(\frac{p_r}{p_c}\right)$. Since j must be no more than n, we have $j = min\left(n, \frac{p_r}{p_c}\right)$.

3.1.1.3 Timeout Loss Indications

In this subsection we model the throughput of cumulative flow for loss indications that are of type "time out" (TO). As already mentioned, TCP's throughput collapse in many-to-one synchronized communication is mainly caused by two kinds of time-outs, namely, Intermediate Block Transfer Timeouts (IBTT) and Anterior Block Transfer Timeouts (ABTT).

Figure 3.3 shows the scenario where IBTT happens in ns-2 [44, 45] simulations. The simulation consists of four senders that transmit synchronized data block to the same receiver. As with the standard Incast communication pattern, the client makes a request for the next block only when the previous block has been completely received. The advertised window size of the receiver is set to 1,00 packets, which is large enough to have no impact on the congestion window evolution at the sender. Figure 3.3 plots the window evolution of three of the four senders involved. The dotted vertical lines running across all three evolutionary graphs indicate the completion of a block transfer. We can see that at time $t \approx 13.559886s$, the client successfully receives block number 20. Following the complete reception of the block, the client makes a request for the transfer of the next block and all senders start transmitting their share for block 21. During transfer of this block, sender 1 at time $t = 13.563974s$, experiences a TO. Since the loss indicator is a timeout, sender 1 waits for a period of time T_0, defined by TCP's retransmission timer before retransmitting its lost packets. And although the other servers involved in the block transfer complete transmitting their share of the block well before the recovery of sender 1, the client does not make a request for a new block till sender 1 also follows suit. Hence, the shared link is completely idle between $13.568913s \sim 13.764448s$, which results in throughput collapse. By observing the congestion window evolution of sender 1, we find that although the packets in its congestion window at $t \approx 13.559886s$, were all successfully transmitted, the server received less than 3 duplicate ACKs resulting in a TO.

Figure 3.4 illustrates the situation where ABTTs occur. In this ns-2 simulation setup, ten senders transmit synchronized data block to the same receiver. As with the simulations for IBTT, the advertised window size of the receiver is set to 1,00 packets, which is again large enough to have no impact on the congestion window evolution at the sender. Figure 3.4 plots the window evolution of three of the ten senders involved. Like before, the dotted vertical lines running across all three evolutionary graphs indicate the completion of a block transfer. Here, we notice that sender 10 experiences a TO very early ($t \approx 1.855538s$) in the transfer of block 9 to the receiver. By the time sender 10 resumes with its transmission (at $t \approx 2.054982s$), all the remaining servers involved, have finished transmitting their share of the data and are waiting for sender 10 to catch up. Like with IBTT, the shared link remains completely idle during this interval ($1.875551s \sim 2.054982s$), which drastically reduces the overall throughput of the Incast traffic. However once sender 10 resumes its transmission, it does not have to compete with any other sender for a portion of the shared bandwidth. This results in a large congestion window for sender 10 at the end of the transfer of block 9. At the beginning of the next block transfer, all senders start off by injecting

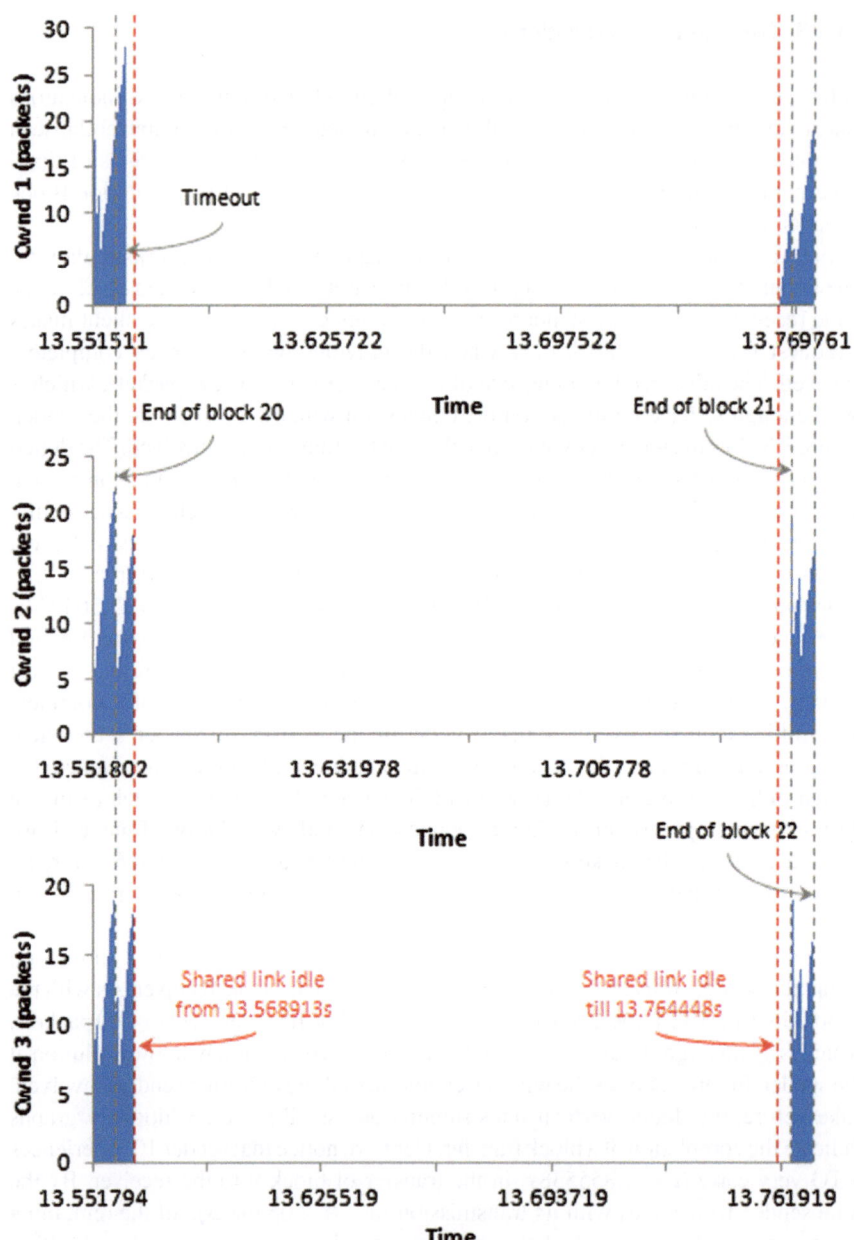

Fig. 3.3 Scenario for intermediate block transfer timeouts

their whole windows into the network. The small buffers at the intermediate Ethernet switch are easily overwhelmed by large windows of senders like 10 and as a result, a lot of packets get dropped. Unfortunately, few senders like sender 9, lose all the packets in their congestion window resulting in an early TO for block 10. And like sender 10 during transfer of block 9, sender 9 too ends up with a large congestion window during the transfer of block 10. The cycle repeats for block 11 too, with sender 8 experiencing an early TO.

Through investigating numerous simulations, we find that IBTT dominates TCP throughput when n is small, while ABTT dominates Incast when n is large.

Consider the evolution of the cumulative window W, in the presence of loss indications that include type "TO", as shown in Fig. 3.5. Timeouts occur when any composite flow f loses packets (or ACKs) and receives less than three duplicate ACKs in response. The loss experiencing flow then waits for a period of time denoted by T_0 before retransmitting its non-acknowledged packets. Following a timeout, the congestion window of the flow W_f is reduced to one, and only one packet is thus resent in the first round after the timeout. In case the composite flow suffers another timeout before successfully retransmitting the packets lost during the first timeout, the period of timeout doubles to $2T_0$; this doubling is repeated for each unsuccessful retransmission until $64T_0$ is reached, after which the timeout period remains constant at $64T_0$ [36].

The evolution of the cumulative window W depicted in Fig. 3.5 is an approximation of the real Incast traffic pattern observed during timeouts. Because we have assumed all n flows to be synchronized in terms of rounds, when one composite flow experiences a timeout, the remaining flows refrain from transmitting data as well. However, in the real world when one composite flow experiences a loss, the other $(n-1)$ flows continue to transmit their remaining share of data (e.g. Figs. 3.3 and 3.4).

Slow Start is another aspect of TCP that we have conveniently chosen to ignore in our handling of TO type loss indicators. Following a timeout, TCP uses a mechanism called "Slow Start" to increase its congestion window. Slow Start operates by observing that the rate at which new packets should be injected into the network is the rate at which the acknowledgments are returned by the other end. Unlike the Congestion Avoidance phase where the congestion window is increased by one segment per round trip time, the Slow Start increases the congestion window by one segment for every ACK received. This provides for an exponential growth of the congestion window after it was reduced to one following a timeout. The Slow Start phase is usually much shorter than the Congestion Avoidance phase and for the sake of simplicity, we choose to ignore this phase in our model of Incast.

Despite these aforementioned approximations, we believe that we have still managed to capture the essential aspects of the Incast phenomenon, as indicated by the generally good fit between our model and the simulations as discussed in Sect. 3.2.

Let Z_i^{TO} denote the duration of a sequence of timeouts and Z_i^{TD} denote the time interval between two consecutive timeout sequences. We define S_i to be

$$S_i = Z_i^{TD} + Z_i^{TO} \tag{3.22}$$

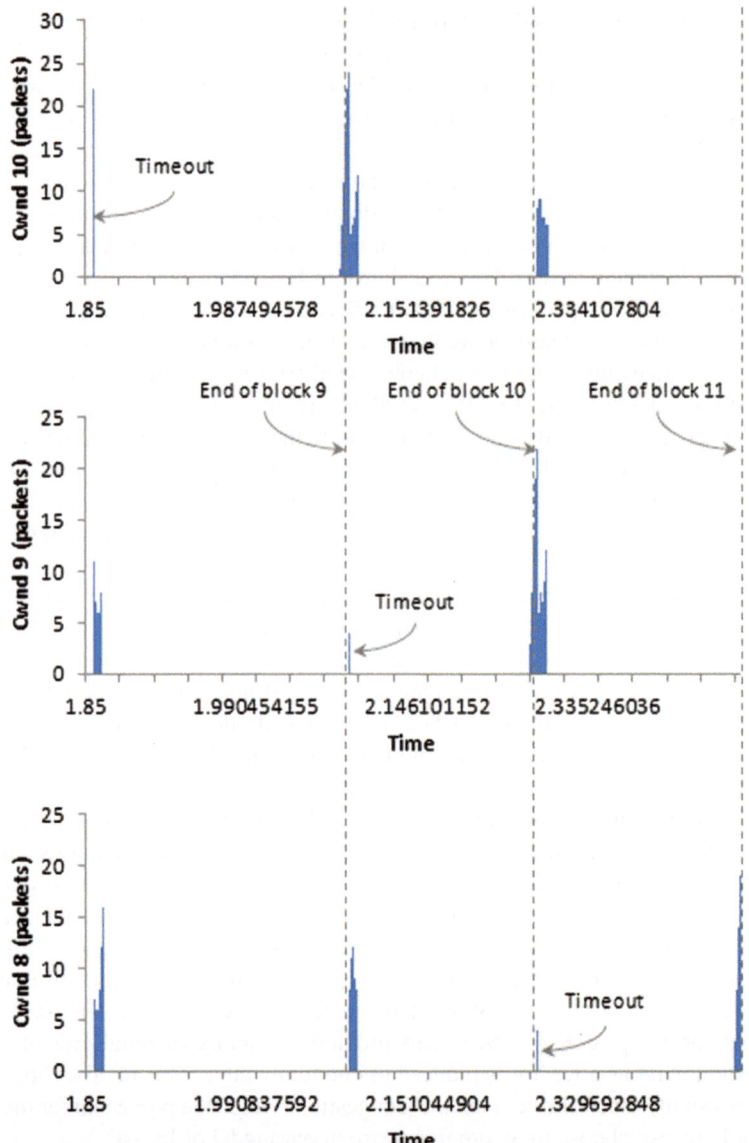

Fig. 3.4 Scenario for anterior block transfer timeouts

Let M_i be the number of packets sent during S_i. Then $\{(S_i, M_i)\}_i$ is an i.i.d. sequence of random variables [36] from which we have,

$$B = \frac{E[M]}{E[S]} \qquad (3.23)$$

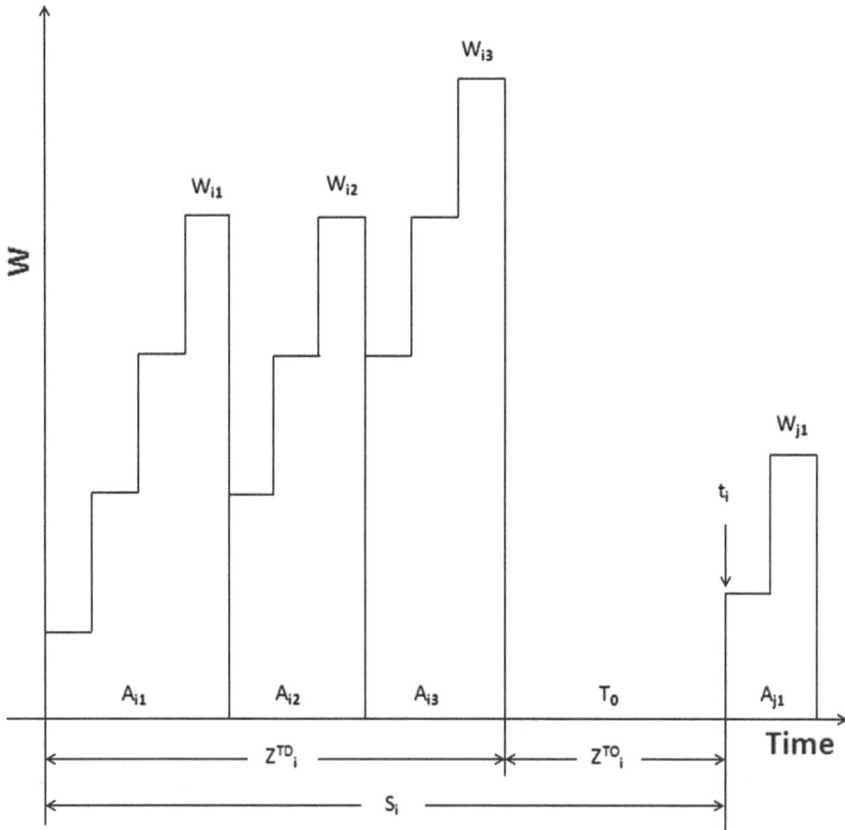

Fig. 3.5 Evolution of W over time when loss indications are TD and TO

Let v_i be the number of TD periods in interval Z_i^{TD}. For the jth TD period of interval Z_i^{TD}, we define Y_{ij} to be the number of packets sent in the period, A_{ij} to be the duration of the period, X_{ij} to be the number of rounds in the period, and W_{ij} to be the cumulative window size of n parallel synchronized TCP flows at the end of the period. From these definitions we have,

$$M_i = \sum_{j=1}^{v_i} Y_{ij} \tag{3.24}$$

and,

$$S_i = \sum_{j=1}^{v_i} A_{ij} + Z_i^{TO} \tag{3.25}$$

Thus,

$$E[M] = E\left[\sum_{j=1}^{v_i} Y_{ij}\right] \qquad (3.26)$$

and,

$$E[S] = E\left[\sum_{j=1}^{v_i} A_{ij}\right] + E\left[Z_i^{TO}\right] \qquad (3.27)$$

If we assume $\{v_i\}_i$ to be an i.i.d. sequence of random variables, independent of Y_{ij} and A_{ij} [36], then we have

$$E\left[\left(\sum_{j=1}^{v_i} Y_{ij}\right)_i\right] = E[v]E[Y] \qquad (3.28)$$

and,

$$E\left[\left(\sum_{j=1}^{v_i} A_{ij}\right)_i\right] = E[v]E[A] \qquad (3.29)$$

To derive $E[v]$ observe that, during Z_i^{TD} the time between two consecutive timeout sequences, there are v_i TDPs, where each of the first $(v_i - 1)$ end in a TD, and the last TDP ends in a TO. It follows that in Z_i^{TD} there is one TO out of v_i loss indications. Therefore if we denote by Q the probability that a loss indication ending a TDP is a TO, we have $Q = \left(\frac{1}{E[v]}\right)$. Consequently,

$$B = \frac{E[Y]}{E[A] + Q \times E[Z^{TO}]} \qquad (3.30)$$

Since A_{ij} and Y_{ij} do not depend on timeouts, their means are those derived in 3.3 and 3.10. To compute throughput of n parallel synchronized TCP connections using 3.30 we must still determine Q and $E[Z^{TO}]$.

We begin by deriving an expression for Q. Let j be the mean number of composite flows experiencing packet loss at the end of a TDP as discussed in the previous subsection. For simplicity, we assume that at most, only one "TO" type loss indication occurs at the end of a TDP. That is, of the j composite flows that lose packets at the end of a TDP, no more than one flow experiences a timeout event. Since a timeout is either of type IBTT or of type ABTT, the probability of a TO type loss indication ending a TDP can be expressed as,

$$Q = min\left(1, Q_{ibtt} + Q_{abtt}\right) \qquad (3.31)$$

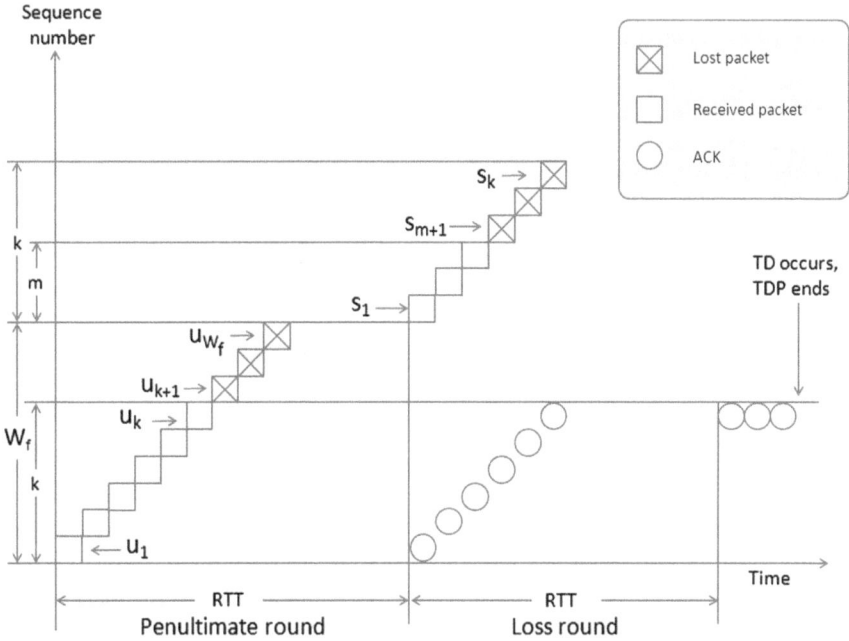

Fig. 3.6 Packet and ACK transmissions preceding a loss indication

where Q_{ibtt} and Q_{abtt} are probabilities of ending a TDP with a single timeout indication of type IBTT and ABTT respectively.

Next, we focus on deriving an expression for Q_{ibtt}—the probability of a composite flow experiencing an IBTT at the end of a TDP. Consider the round where a composite flow f, loses its packets; we will refer to this round as the "penultimate" round (see Fig. 3.6). Let W_f be the size of the flow's congestion window. Thus packets $u_1, \ldots, u_{W_f}$ are sent in the penultimate round. Packets $u_1, \ldots, u_k$ are acknowledged and packet u_{k+1} is the first one to be lost. Since we have assumed packet losses within a round to be correlated, if a packet is lost all packets that follow it till the end of the round are also lost. Thus, all packets following u_{k+1} in the penultimate round are also lost. However, since packets $u_1, \ldots, u_k$ are ACKed, another k packets, $s_1, \ldots, s_k$ are sent in the next round, which we will refer to as the "loss" round. This round of packets may have another loss, say packet s_{m+1}. Again, our assumptions on packet loss correlation mandates that packets $s_{m+2}, \ldots, s_k$ are also lost in the loss round. The m packets successfully sent in the loss round are responded to by ACKs for packet u_k, which are counted as duplicate ACKs. If the number of such ACKs is higher than three, then a TD indication occurs, otherwise an IBTT occurs. In both cases, the current period between losses, TDP, ends. We denote by $A(w, k)$ the probability that the first k packets are ACKed in a round of w packets, given there is a sequence of one or more losses in the round. Then,

$$A(w, k) = \frac{(1 - p_r)^k p_r}{1 - (1 - p_r)^w} \tag{3.32}$$

Also, we define $C(g, m)$ to be the probability that m packets are ACKed in sequence in the loss round (where g packets were sent) and the rest of the packets in the round, if any are lost. Then,

$$C(g, m) = \begin{cases} (1 - p_r)^m p_r, & m < g \\ (1 - p_r)^n, & m = g \end{cases} \tag{3.33}$$

Then, $\hat{Q}_{ibtt}(w)$, the probability that a loss in a congestion window of size w is an IBTT, is given by,

$$\hat{Q}_{ibtt}(w) = \begin{cases} 1, & \text{if } w \leq 3 \\ \sum_{k=0}^{2} A(w, k) + \sum_{k=3}^{w} \left(A(w, k) \times \sum_{m=0}^{2} C(k, m) \right), & \text{otherwise} \end{cases} \tag{3.34}$$

Since an IBTT occurs if the number of packets successfully transmitted in the penultimate round, k, is less than three or if the number of packets successfully transmitted in the loss round, m is less than three. Also, due to the assumption that packets following s_{k+1} are lost independently of packets following u_{k+1} (since they occur in different rounds), the probability that there is a loss at u_{k+1} in the penultimate round followed by a loss at s_{m+1} in the loss round equals $A(w, k) \times C(k, m)$.

Therefore, Q_{ibtt}, the probability that composite flow's loss indication is an IBTT, can be expressed as

$$Q_{ibtt} = \sum_{w=1}^{\infty} \hat{Q}_{ibtt}(w) P[W_f = w] = E\left[\hat{Q}_{ibtt}\right] \tag{3.35}$$

We can approximate this to,

$$Q_{ibtt} \approx \hat{Q}_{ibtt}(E[W_f]) \tag{3.36}$$

where $E[W_f]$ is the mean congestion window size of a composite flow, derived from the Eq. 3.12.

To begin deriving an expression for Q_{abtt} we must first consider the number of packets transmitted in a TDP in relation to the size of the block being transferred. Let L be the size of the block that all n senders are trying to transmit to the destination. If $E[Y]$ is the mean number of packets sent during a TD-period (Eq. 3.10), the average number of TDPs needed to transfer a block of size L, can be expressed as,

$$\rho = \frac{L}{E[Y]} \tag{3.37}$$

If δ is the mean number of ABTTs occurring at the start of a block transfer, the series δ_i and ρ_i can be assumed to be mutually independent sequence of i.i.d. random variables from which, the probability of a TDP ending due to a TO indication of type ABTT can be expressed as

$$Q_{abtt} = \frac{E[\delta]}{E[\rho]} \tag{3.38}$$

We can substitute the results of Eqs. 3.36 and 3.38 in 3.31 to get an expression for Q—the probability that a loss indication ending a TDP is a TO.

Next, we consider the derivation of $E[Z^{TO}]$, the average duration of a timeout sequence. Since we have assumed that there can be at most one timeout at the end of a TDP, we can approximate $E[Z^{TO}]$ with T_0.

By substituting the obtained expressions for Q and $E[Z^{TO}]$ into Eq. 3.30, we now obtain the following expression for B

$$B = \left(\frac{E[Y]}{RTT \times E[X] + Q \times E[Z^{TO}]} \right) \quad \text{where,} \tag{3.39}$$

$$E[Y] = \left(\frac{1}{p_c} \right)$$

$$E[X] = \left(\frac{2j^2 p_c - np_c j + \sqrt{n^2 p_c^2 j^2 - 4n p_c^2 j^3 + 4j^4 p_c^2 + 24n^2 p_c j}}{6n^2 p_c} \right)$$

$$j \approx min\left(n, \frac{p_r}{p_c} \right)$$

$$Q = min(1, Q_{ibtt} + Q_{abtt})$$

$$Q_{ibtt} \approx \hat{Q}_{ibtt}(E[W_f])$$

$$E[W_f] = \left(\frac{2n}{j} \times \frac{2j^2 p_c - np_c j + \sqrt{n^2 p_c^2 j^2 - 4n p_c^2 j^3 + 4j^4 p_c^2 + 24n^2 p_c j}}{6n^2 p_c} \right)$$

$$\hat{Q}_{ibtt}(w) = \begin{cases} 1, & \text{if } w \leq 3 \\ \sum_{k=0}^{2} A(w,k) + \sum_{k=3}^{w} \left(A(w,k) \times \sum_{m=0}^{2} C(k,m) \right), & \text{otherwise} \end{cases}$$

$$A(w,k) = \frac{(1-p_r)^k p_r}{1-(1-p_r)^w}$$

$$C(k,m) = \begin{cases} (1-p_r)^m p_r, & m < k \\ (1-p_r)^n, & m = k \end{cases}$$

$$Q_{abtt} = \frac{E[\delta]}{E[\rho]}$$

$$E[Z^{TO}] = T_0$$

In Sect. 3.2, we verify whether the Eq. 3.39 successfully models the behavior of Incast or not. Henceforth we will refer to the model expressed in Eq. 3.39 as the "Full Model".

3.2 Validation and Analysis

In this section, we validate the performance of our Incast model using the ns-2 simulator. With simulations we demonstrate that the throughput expression derived in the previous section works relatively well for broad range of conditions.

For our ns-2 simulations, we used the topology depicted in Fig. 3.7 which is commonly used to study a set of parallel, synchronized flows sharing the same bottleneck link. We vary various parameters like, number of flows, block size as well as buffer length to validate our model under different experimental conditions.

Our ns-2 simulation configuration depicted in Fig. 3.7 consists of a cluster based storage system where storage client and storage servers are all connected to the same switch. In this environment, data blocks are striped over multiple servers, such that each server stores a fragment of the data block denoted as the Server Request Unit (SRU) in Fig. 3.7. A client requesting a data block sends request packets to all storage servers that contain SRUs for that particular block; the client requests the next block only after it has received all the data for the current requested block. That is, if the client requests a data block from n servers, it sends request for the next block only after receiving ($n \times SRU$) bytes of data in total.

Next, we measure the throughput of n parallel, synchronized TCP flows at the shared bottleneck link after varying the number of storage servers involved in data transfer. To more accurately model the real-world scheduling variance, we also add a random scheduling delay of up to 20 μs between every consecutive data request from the client. Table 3.1 lists various other parameters that were used in our experiments. Notice that we have enabled "Slow start" in our experiments even after choosing to ignore it for our model. As we demonstrate later in this section, the impact of "Slow

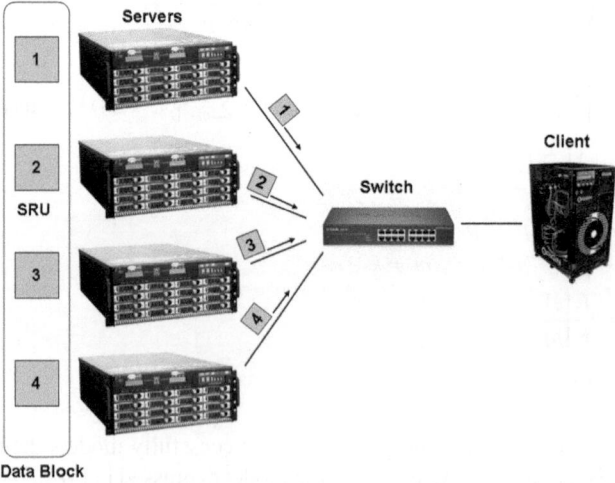

Fig. 3.7 Setup for n parallel, synchronized TCP flows sharing a bottleneck

Table 3.1 Simulation parameters with default settings

Parameter	Default
Number of servers	–
SRU size (L)	256 KB
Link capacity (C)	1 Gbps
Link delay (D)	50 μs
Switch buffer size (B)	32 KB
Segment size (S)	1 KB
TCP implementation	NewReno
Receive window size	1,000 segments
Duplicate ACK threshold	3
Slow start	enabled
RTO$_{min}$	200 ms

start" on Incast is negligible; our model produces a good fit with the simulations despite ignoring "Slow start". Each trial in the experiment runs for 40 sec of simulated time, providing enough data transfer to accurately calculate the throughput of the Incast traffic.

Simultaneously, we also gather traces generated by ns-2 for all the traffic simulated in our experiment. Later, we analyze these traces with a set of analysis programs developed by us. These programs compute the values of p_c by dividing the total number of loss indications in the cumulative flow by the total number of packets sent by all flows, p_r by dividing the total number of loss indications in a composite flow by the total number of packets sent by the flow and δ the mean number of ABTTs occurring at the start of a block transfer. Additionally, the programs also measure the round trip time and the average duration of a "single" timeout. These values are then averaged over several runs and our model's throughput computed using Eq. 3.39.

Figure 3.8 compares the throughput of ns-2 Incast simulations to the throughput obtained by our model using Eq. 3.39. From the graph, it can be seen that our model characterizes the general tendency of TCP Incast relatively well, although it underestimates the throughput at the bottleneck link when the number of senders is large.

Figure 3.9 on the other hand compares the throughput of ns-2 Incast simulations to the throughput obtained by our model using Eq. 3.21. It is important to note that the expression in Eq. 3.21 computes the throughput of the Incast traffic when all the composite flows are in congestion avoidance phase and receive only TD type loss indicating events. That is, Eq. 3.21 computes the throughput of the Incast traffic without taking timeouts into account.

Comparing Figs. 3.8 and 3.9 it is clear that, timeouts—both ABTT and IBTT—are essentially the main causes for TCP's throughput collapse under Incast workloads. To better understand the impact of IBTT and ABTT on Incast traffic, we compute the throughput achieved in our model by considering just one type of timeout at a

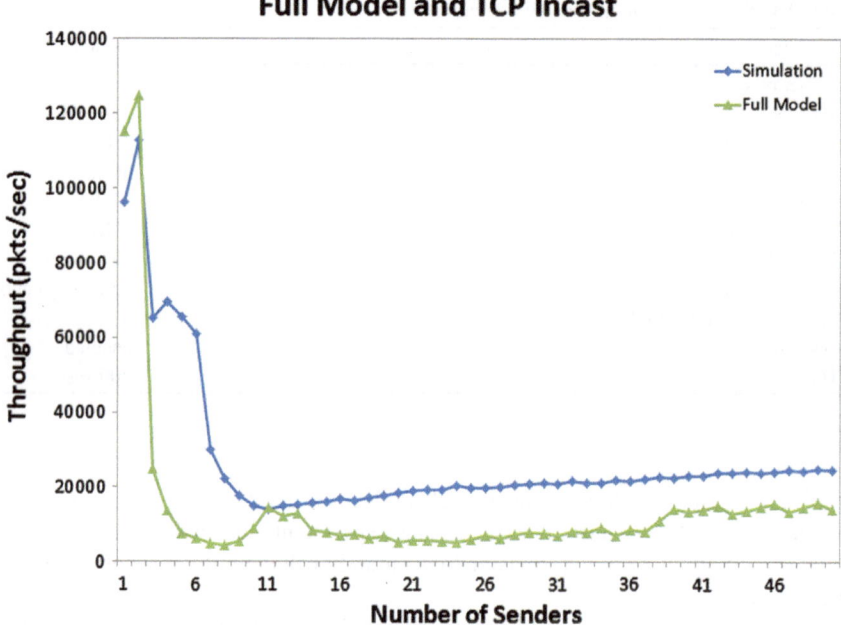

Fig. 3.8 Comparing full model with incast simulation results

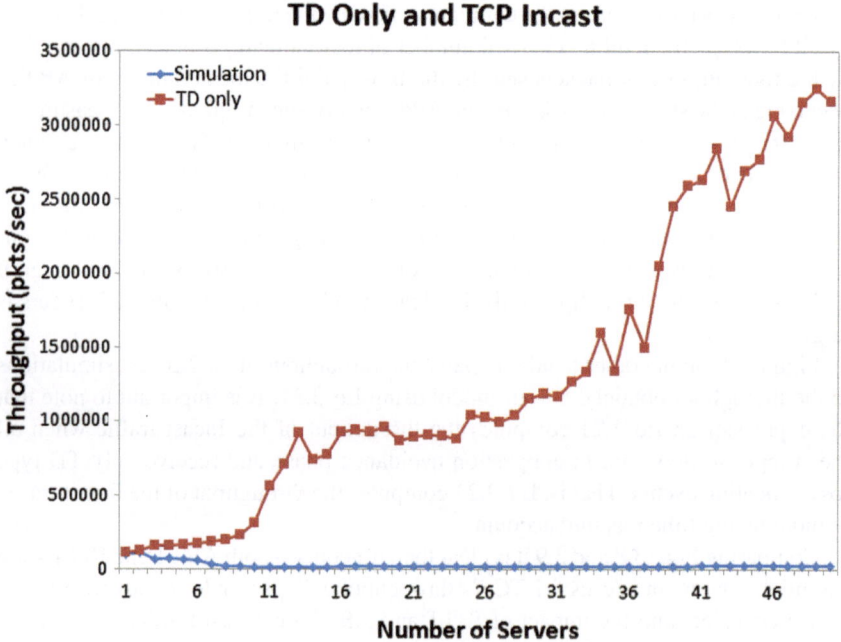

Fig. 3.9 Comparing TD only model with incast simulation results

time. Figures 3.10 and 3.11 plot the throughput resulting from our Full Model when only IBTT and ABTT, are considered respectively.

From Fig. 3.10 it is clear that IBTTs have a greater impact on throughput when the number of senders is small. When the number of senders is between three and eight, our model overestimates the impact of IBTTs when compared to throughput resulting from ns-2 simulations. Also, when the number of senders is greater than eight, we observe that the model's throughput no longer matches that of the ns-2 simulations. This is mainly because the expression for Q_{ibtt} in Eq. 3.36 does not take into account the timeouts happening at the beginning of a block transfer. Furthermore, since Q_{ibtt} in Eq. 3.36 only relies on the probabilities p_c and p_r, even small deviations in their measured values, can result in large fluctuations in the model's throughput.

On the other hand, from Fig. 3.11, it is clear that ABTTs dominate timeouts when the number of senders is large. From the graph, we observe that the ABTTs have little or no impact on the model's throughput when the number of senders is less than ten. As the number of senders increase, some of them finish transmitting their SRUs early allowing the remaining senders to transmit their SRUs using the additional bandwidth vacated by the finished peers. This results in large transmission windows for some of the senders at the end of the block transfer. At the beginning of the next block transfer, all senders begin by injecting their entire congestion windows into the network. With some senders injecting larger number of segments, this packet burst

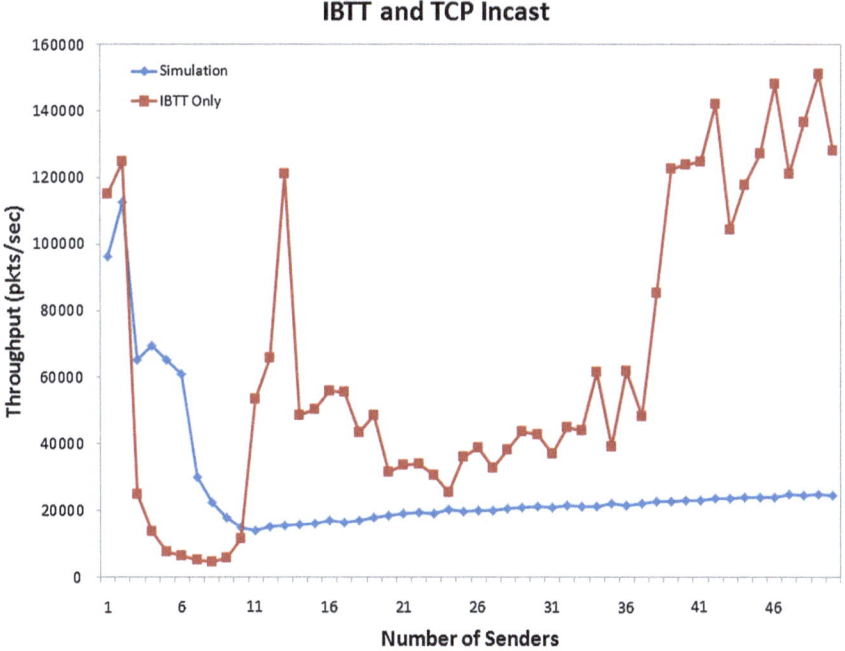

Fig. 3.10 Impact of IBTT on proposed model

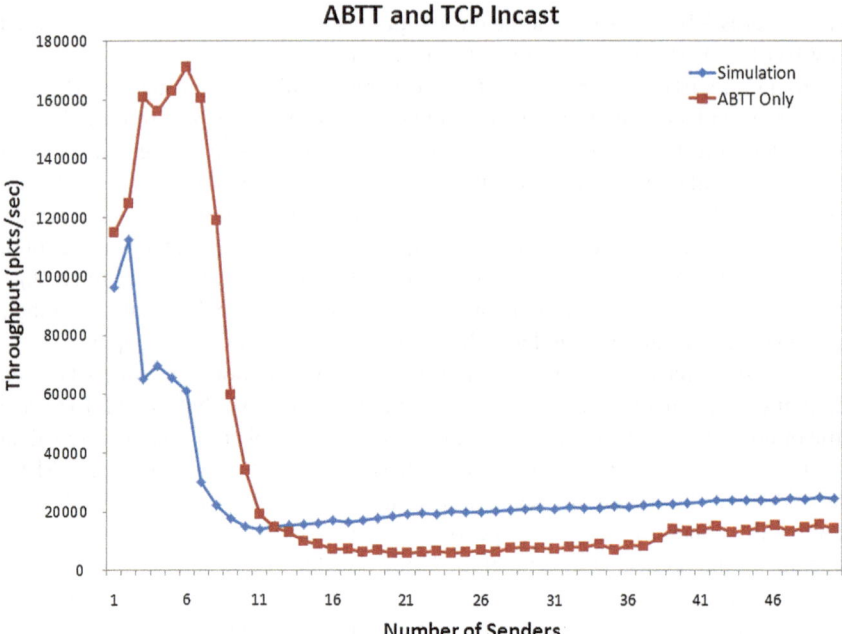

Fig. 3.11 Impact of ABTT on proposed model

at the beginning of a block transfer overwhelms the bottleneck link's port buffers resulting in packet drops and ABTTs.

It is interesting to observe that ABTTs are the result of the start-stop nature of the synchronized block transfers. The senders stop transmission after completing their SRU transfer and start transmitting again only after receiving a new transfer request. This new transfer request results in a packet burst which floods the buffers at the bottleneck link resulting in packet loss and ABTTs. If the senders each had SRUs of infinite size like in [36], there would be no start-stop pattern to Incast's traffic and hence, no ABTTs. This would have resulted in the senders experiencing only IBTTs in which case, the expression for Q_{ibtt} in Eq. 3.36 would have been sufficient for estimating the probability of a timeout at the end of a TDP.

Going back to Fig. 3.8, we can now analyze the performance of our model in two parts: the first part, where the number of senders is large and ABTTs have a bigger impact and the second part, where the number of senders is small and IBTTs are dominant. In the first part, it is clear that our model underestimates the throughput of multiple TCP flows at the bottleneck link. This is because our model overestimates the time spent in recovering from an ABTT. If we were to revisit the expression for Q_{abtt} in Eq. 3.38, we find that δ is defined as the mean number of timeouts occurring at the beginning of a block transfer. While our model simply counts the average number of flows experiencing timeouts at the beginning of a block transfer, from the traces generated by ns-2, we find that most of these timeouts occur simultaneously. With simultaneous timeouts, the participating flows wait for a single T_0 period before recovering, although the timeout events are counted separately. Since we do not take simultaneous timeouts into account while deriving an expression for Q_{abtt}, the estimated duration between two successive TDs in our model is slightly longer than that of ns-2. This in turn decreases the number of packets estimated per unit time, which is why our model underestimates Incast throughput when the number of senders involved is large.

In the second part of our performance analysis of the Full Model, we find that the model predicts a huge drop in Incast throughput when the number of senders is approximately three. The throughput obtained via ns-2 simulations on the other hand, appears to have a step around the four senders mark, followed by a significant drop in performance when the number of senders is around seven. In order to better understand this discrepancy in the results, we analyzed the traces generated by ns-2 simulations in great detail. From these traces we found that whenever the number of senders is less than or equal to three, IBTTs are caused by only one reason—whole window losses. That is, when the sender experiences a timeout with $n \leq 3$, it loses all the packets in its congestion window, without receiving a single ACK in return. This type of loss happens when two or more individual flows simultaneously attempt to fill the bottleneck link buffer, resulting in at least one flow losing all its packets. On the other hand when the number of senders is greater than three, IBTTs in ns-2 simulations happen only because of one reason—ack of enough duplicate ACKs. This type of loss happens when a sender loses packets in both "penultimate" as well as "loss" rounds due to severe congestion at the bottleneck link, as discussed earlier in Sect. 3.1 while deriving an expression for $C(g, m)$ in Eq. 3.33.

Taking into account the exclusive nature of IBTT type of timeouts in ns-2 simulations, we can now compute the following new expression for $\hat{Q}_{ibtt}(w)$.

$$
\hat{Q}_{ibtt}(w) = \begin{cases} 1, & \text{if } w \leq 3 \\ A(w, 0), & \text{if } n \leq n^* \\ \sum_{k=3}^{w} \left(A(w, k) \times \sum_{m=0}^{2} C(k, m) \right), & \text{otherwise} \end{cases} \tag{3.40}
$$

where n^* is the number of senders after which IBTTs are entirely caused by insufficient duplicate ACKs.

By substituting the Eq. 3.40 in Eqs. 3.35 and 3.36, we end up with a new expression for B, the throughput of the Incast traffic across the bottleneck link. We refer to the model resulting from Eq. 3.40 as the "Split Model" as opposed to the "Full Model" derived in Eq. 3.39.

Figure 3.12 compares the throughput of ns-2 Incast simulations to the throughput obtained by the Split Model described above. From the graph it can be seen that the Split Model is much better at characterizing the overall tendency of TCP Incast. When the number of senders n is less than or equal to three, Split Model only considers whole window losses for IBTTs. Beyond that, as the number of senders increase, Split Model only considers insufficient duplicate ACKs for IBTTs. When combined with timeouts of type ABTT resulting from packet burst at the start of a block transfer,

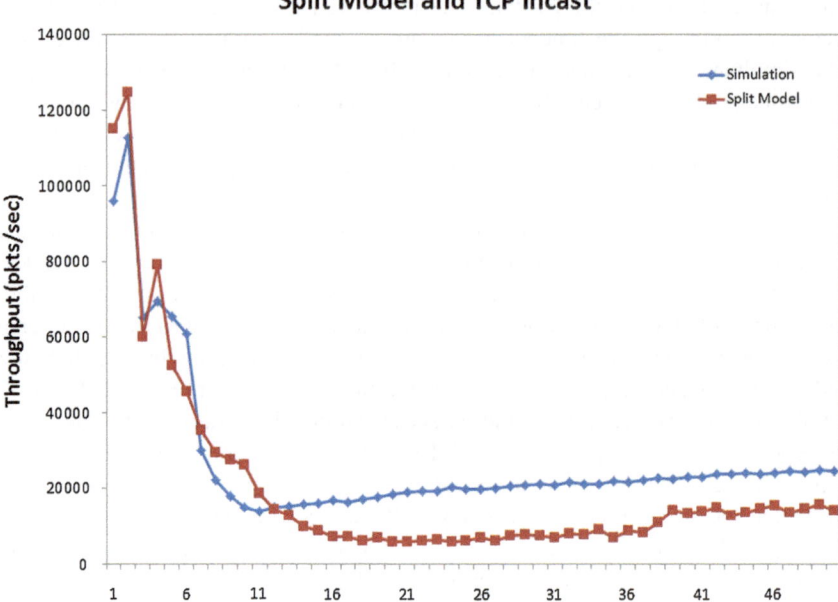

Fig. 3.12 Comparing split model with incast simulation results

the Split Model appears to model the Incast traffic much better than our earlier Full Model.

Next, we compare the performance of our proposed models against the simulation results for different sizes of switch buffers. From the four simulation curves in Figs. 3.13, 3.14, 3.15 and 3.16, we can summarize the following:

- Larger switch buffer improves the throughput at the bottleneck link with different number of senders n. This can be explained by our proposed model. Larger buffer size implies fewer dropped packets i.e., smaller values for probabilities p_c and p_r. Hence, the expected number of packets Y in a TDP increases.
- Larger switch buffer shifts throughput collapse to the right. That is, for larger switch buffers, several parallel, synchronized senders can transmit data without experiencing Incast. This is because larger switch buffers can cache more packets, thereby reducing the probability of packet loss. And since packet losses lead to Anterior Block Transfer Timeouts as the number of senders increase, large buffers delay the onset of performance loss by reducing the number of ABBTs.

Following this, we compare the performance of our proposed models against the simulation results for different sizes of SRU at the senders. Figures 3.17, 3.18, 3.19 and 3.20 plot the performance of our proposed model and simulation results for these scenarios. From these graphs, we can summarize the following:

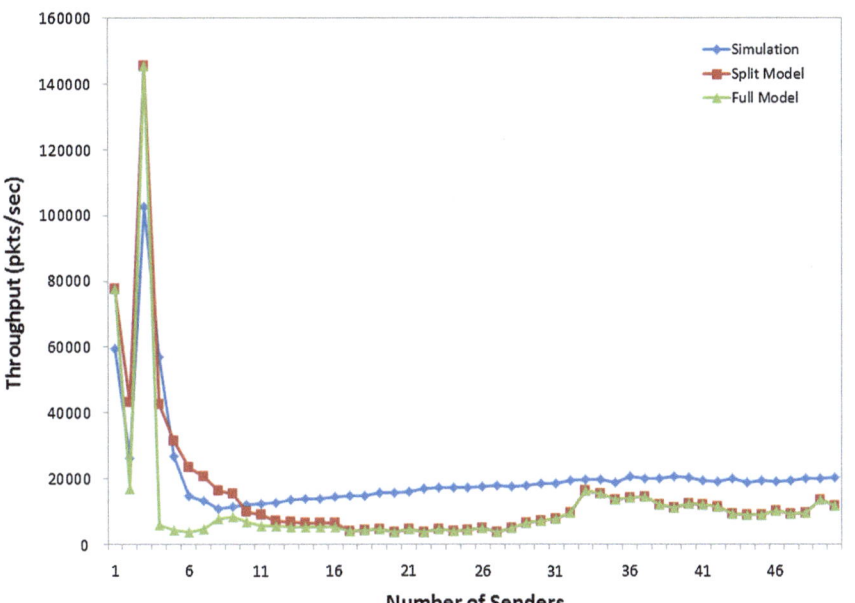

Fig. 3.13 Performance of full model, split model and ns-2 with 16 KB switch buffer

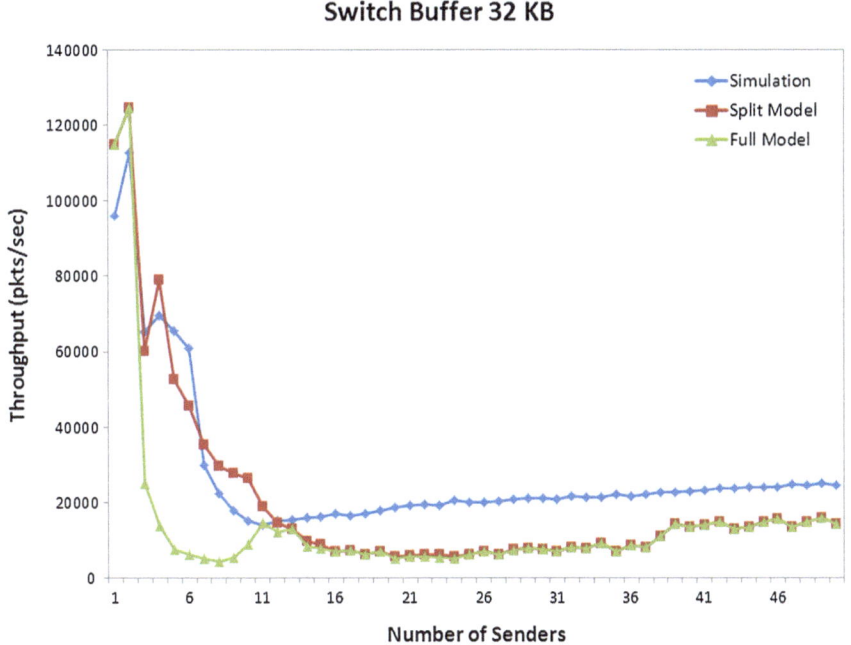

Fig. 3.14 Performance of full model, split model and ns-2 with 32 KB switch buffer

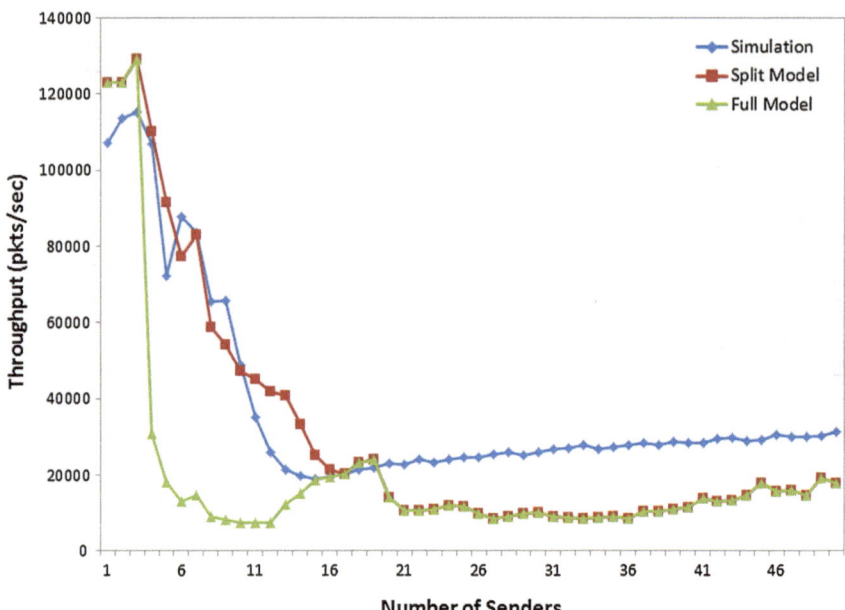

Fig. 3.15 Performance of full model, split model and ns-2 with 64 KB switch buffer

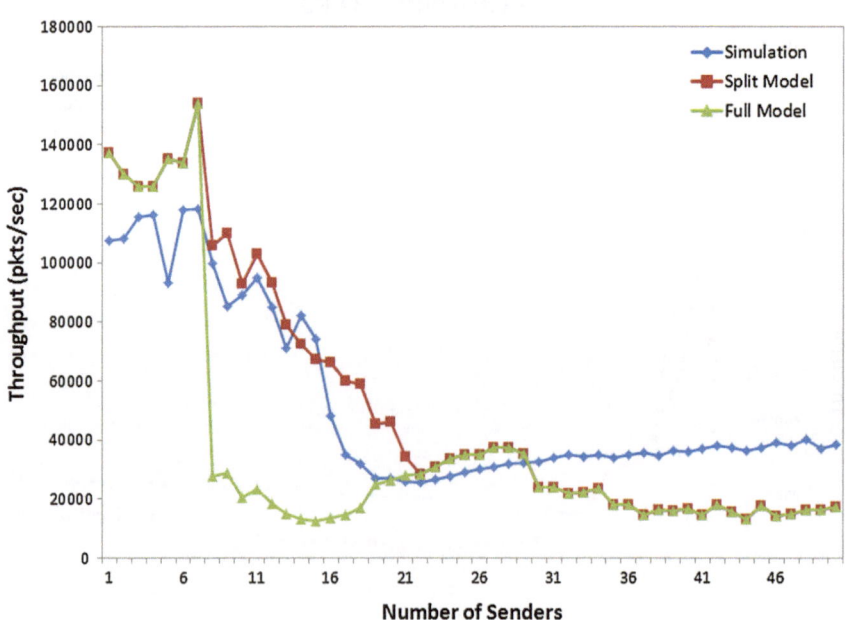

Fig. 3.16 Performance of full model, split model and ns-2 with 128 KB switch buffer

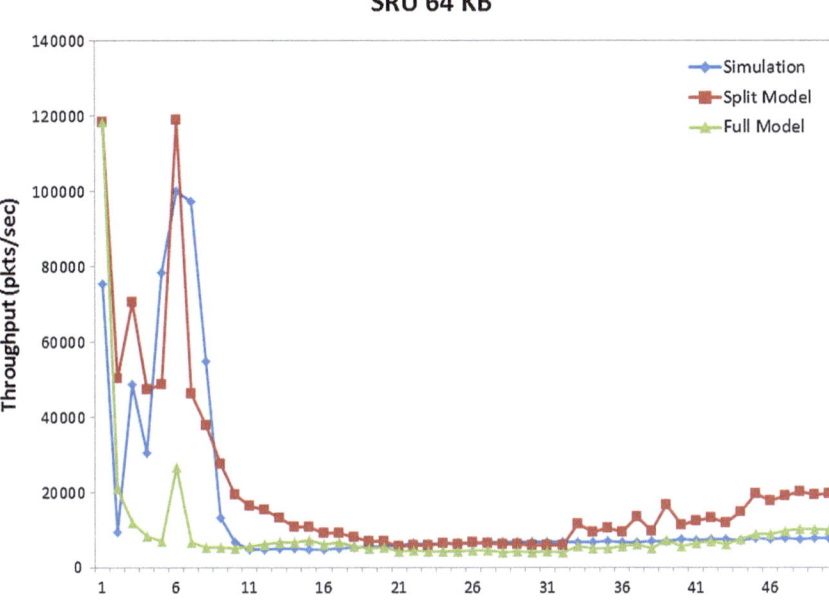

Fig. 3.17 Performance of full model, split model and ns-2 with 64 KB SRU

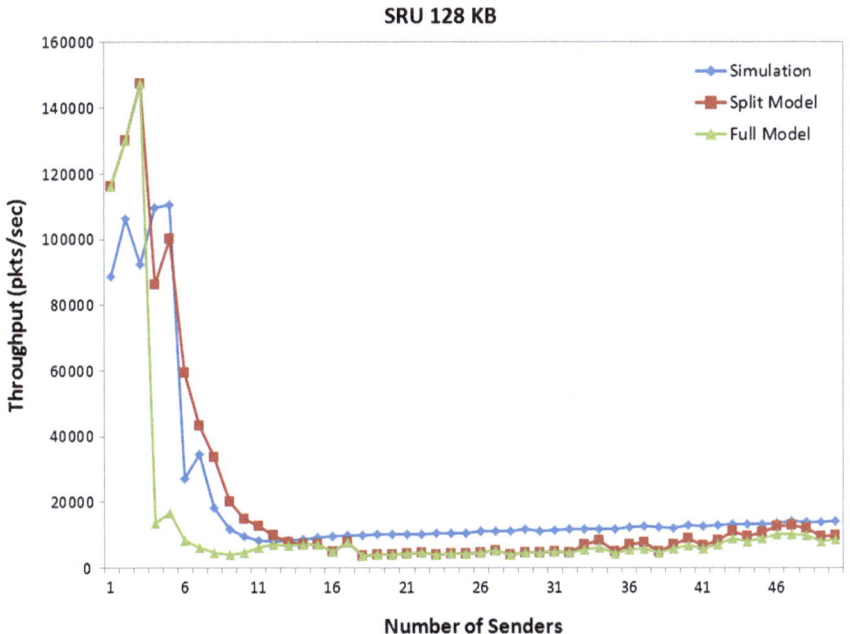

Fig. 3.18 Performance of full model, split model and ns-2 with 128 KB SRU

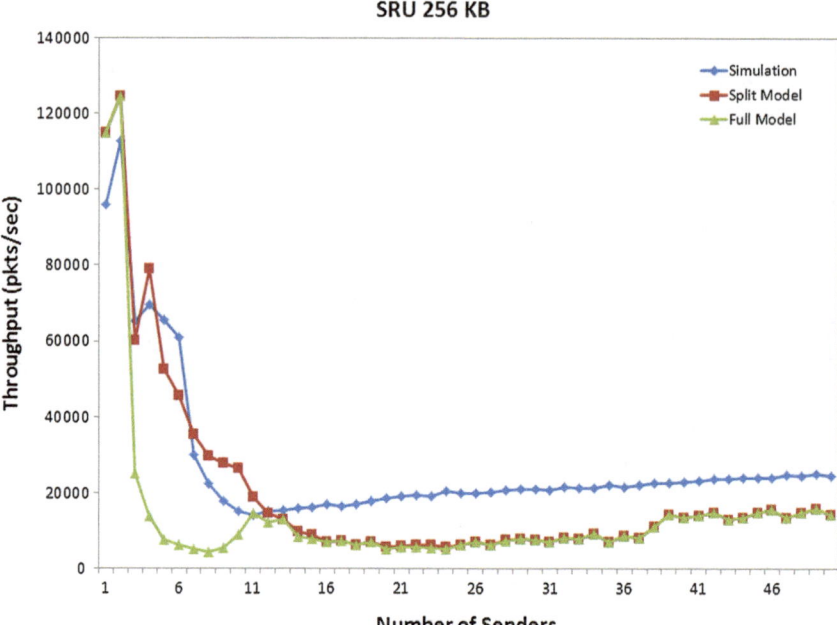

Fig. 3.19 Performance of full model, split model and ns-2 with 256 KB SRU

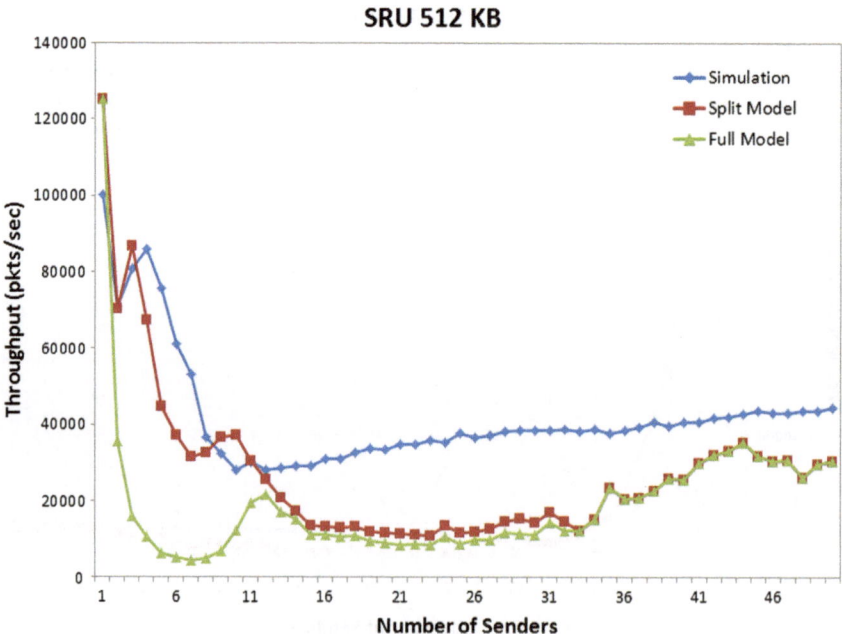

Fig. 3.20 Performance of full model, split model and ns-2 with 512 KB SRU

- We can see that the throughput increases when the SRUs grow larger. But large SRU size has little impact on the onset of throughput collapse. According to our model, SRU size is irrelevant to the maximum cumulative window size.
- With larger SRU size, the time wasted by a TO period to the time spent by unlucky flows transmitting packets becomes smaller. As a result, the throughput across the bottleneck link increases after Incast.

3.2.1 Comparing with Single Flow Model

Due to the simplicity of our model, it is tempting to believe that a similar result could also be obtained by simply multiplying the original equation from [36] with the number of flows n. In Fig. 3.21, we compare the simulations results from ns-2 with $n \times$ (*equation from* [36]). Here, in order to obtain the curve for the expression $n \times$ (*equation from* [36]), we have substituted the packet loss probability p in [36] with average p_r from Sect. 3.1, the probability that a packet (belonging to any flow) is lost.

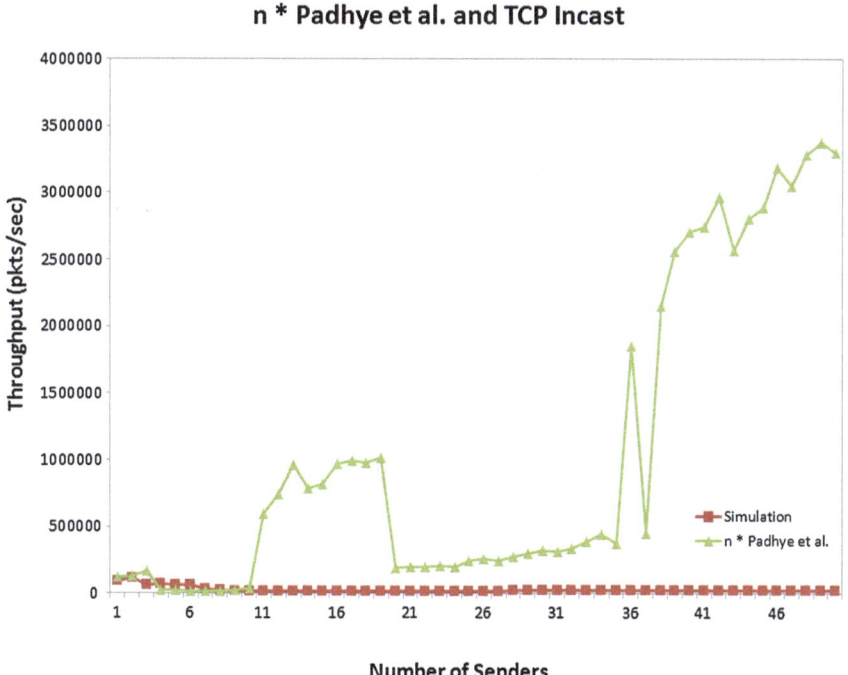

Fig. 3.21 Comparing n * (equation from Padhye et al.) with incast simulation results

As expected the expression $n \times$ [36]), works reasonably well for small values of n. That is because for smaller values of n, the resulting TCP timeouts are largely dominated by IBTTs. On the other hand, when n is reasonably large, ABTTs happen more frequently. Unfortunately, the equation in [36] does not take ABTTs into account. Hence, as n becomes large, the predicted throughput of the expression $n \times$ (*equation from* [36]) is orders of magnitude larger than the one obtained through simulations. This discrepancy in performance only reinforces our decision to develop a new analytical model to examine various aspects of Incast.

3.3 Summary

In this chapter, we built analytical models to understand the essential causes behind TCP Incast, which is an important problem in data center networks. Existing investigations on TCP Incast try to find a good solution to the problem despite incurring high costs. For example, some of the prevalent Incast solutions include, substituting TCP with UDP, reducing RTO_{min} value, increasing switch buffer size, limiting the number of senders in Incast transfers, etc... To solve TCP Incast substantially, the fundamental reasons behind it should be first explored. Unfortunately, almost all existing studies of TCP, model the protocol considering a single flow with an application that has an infinite amount of data to transmit. Furthermore, there are practically no prior TCP models that study the protocol's performance under synchronized traffic workloads in high speed, low latency, data center networking environments. Our models fill this void by extending the single flow model in [36] to multiple synchronized flows, where each flow contributes a finite amount of data.

In our work, we find that two types of timeouts, ABTT and IBTT, are together responsible for TCP's throughput collapse in many-to-one synchronized traffic workloads. IBTT, which is caused by one of the last three packets in a round being dropped, has a greater impact on throughput when the number of senders is small. ABTT, which is caused by the start-stop nature of Incast traffic at the beginning of a block transfer, dominates timeouts when the number of senders is large. We validate the performance of our proposed models by comparing them with simulation data. Although our models characterize the overall effect of Incast pretty well, we find them to be a little conservative in their estimation of the cumulative throughput. This is because, our models overestimate the frequency of ABTTs, resulting in longer delays and lower throughput when compared to simulation data.

From our experiments we were also able to demonstrate that larger switch buffers can not only improve the throughput of the Incast traffic but can even delay the onset of throughput collapse. Similarly, we show that larger SRUs can also improve the throughput of Incast traffic. However, we find that the size of the SRU has little impact on the onset of throughput collapse.

Finally, all the insights gained in building and validating our proposed models, will help us develop more effective solutions that address the problem of TCP Incast, preferably at lower costs.

References

1. T. Berners-Lee, R. Fielding, H. Frystyk, Hypertext Transfer Protocol—HTTP/1.0, RFC 1945 (Informational), Internet Engineering Task Force, May 1996. http://www.ietf.org/rfc/rfc1945.txt
2. J. Postel, J. Reynolds, File Transfer Protocol, RFC 959 (Standard), Internet Engineering Task Force, Oct. 1985, updated by RFCs 2228, 2640, 2773, 3659, 5797. http://www.ietf.org/rfc/rfc959.txt
3. J. Postel, Simple Mail Transfer Protocol, RFC 821 (Standard), Internet Engineering Task Force, Aug. 1982, obsoleted by RFC 2821. http://www.ietf.org/rfc/rfc821.txt
4. C. Feather, Network News Transfer Protocol (NNTP), RFC 3977 (Proposed Standard), Internet Engineering Task Force, Oct. 2006, updated by RFC 6048. http://www.ietf.org/rfc/rfc3977.txt
5. T. Ylonen, C. Lonvick, The Secure Shell (SSH) Protocol Architecture, RFC 4251 (Proposed Standard), Internet Engineering Task Force, Jan. 2006. http://www.ietf.org/rfc/rfc4251.txt
6. T. Ylonen, C. Lonvick, The Secure Shell (SSH) Authentication Protocol, RFC 4252 (Proposed Standard), Internet Engineering Task Force, Jan. 2006. http://www.ietf.org/rfc/rfc4252.txt
7. T. Ylonen, C. Lonvick, The Secure Shell (SSH) Transport Layer Protocol, RFC 4253 (Proposed Standard), Internet Engineering Task Force, Jan. 2006. http://www.ietf.org/rfc/rfc4253.txt
8. T. Ylonen, C. Lonvick, The Secure Shell (SSH) Connection Protocol, RFC 4254 (Proposed Standard), Internet Engineering Task Force, Jan. 2006. http://www.ietf.org/rfc/rfc4254.txt
9. J. Schlyter, W. Griffin, Using DNS to Securely Publish Secure Shell (SSH) Key Fingerprints, RFC 4255 (Proposed Standard), Internet Engineering Task Force, Jan. 2006. http://www.ietf.org/rfc/rfc4255.txt
10. F. Cusack, M. Forssen, Generic Message Exchange Authentication for the Secure Shell Protocol (SSH), RFC 4256 (Proposed Standard), Internet Engineering Task Force, Jan. 2006. http://www.ietf.org/rfc/rfc4256.txt
11. K. Thompson, G.J. Miller, R. Wilder, Wide-area internet traffic patterns and characteristics, Netwrk. Mag. of Global Internetwkg 11(6), 10–23 (1997). http://dx.doi.org/10.1109/65.642356
12. J. Postel, User Datagram Protocol, RFC 768 (Standard), Internet Engineering Task Force, Aug. 1980. http://www.ietf.org/rfc/rfc768.txt
13. S. McCreary, k. claffy, Trends in wide area IP traffic patterns—a view from Ames internet exchange, in ITC Specialist Seminar, Monterey, Sep 2000
14. R. Chow, Y.-C. Chow, *Distributed Operating Systems and Algorithms* (Addison-Wesley Longman Publishing Co., Inc., Boston, 1997)
15. A. Silberschatz, P.B. Galvin, G. Gagne, *Operating System Concepts*, 6th edn. (John Wiley & Sons, Inc., New York, 2001)
16. G.F. Pfister, *In Search of Clusters*, 2nd edn. (Prentice-Hall, Inc., Upper Saddle River, 1998)
17. R. Buyya, *High Performance Cluster Computing: Architectures and Systems* (Prentice Hall PTR, Upper Saddle River, 1999)
18. D. DeWitt, J. Gray, Parallel database systems: the future of high performance database systems, Commun. ACM 35(6), 85–98 (1992). http://doi.acm.org/10.1145/129888.129894
19. K. Keeton, D. Beyer, E. Brau, A. Merchant, C. Santos, A. Zhang, On the road to recovery: restoring data after disasters, in *Proceedings of the 1st ACM SIGOPS/EuroSys European Conference on Computer Systems 2006*, ser. EuroSys '06. New York: ACM, 2006, pp. 235–248. http://doi.acm.org/10.1145/1217935.1217958
20. K. Keeton, C. Santos, D. Beyer, J. Chase, J. Wilkes, Designing for disasters, in *Proceedings of the 3rd USENIX Conference on File and Storage Technologies*, ser. FAST'04. Berkeley, CA, USA: USENIX Association, 2004, pp. 5–5. http://dl.acm.org/citation.cfm?id=1973374.1973379
21. IEEE Standard for Information Technology—Telecommunications and Information Exchange Between Systems—Local and Metropolitan Area Networks—Specific Requirements—Part 3:

Carrier Sense Multiple Access With Collision Detection (CSMA/CD) Access Method and Physical Layer Specifications, LAN/MAN Standards Committee, New York, 2008. http://standards. ieee.org/about/get/802/802.3.html

22. A. Phanishayee, E. Krevat, V. Vasudevan, D.G. Andersen, G.R. Ganger, G.A. Gibson, S. Seshan, Measurement and analysis of TCP throughput collapse in cluster-based storage systems, in *Proceedings of the 6th USENIX Conference on File and Storage Technologies*, ser. FAST'08. Berkeley: USENIX Association, 2008, pp. 12:1–12:14. http://dl.acm.org/citation. cfm?id=1364813.1364825

23. V. Vasudevan, A. Phanishayee, H. Shah, E. Krevat, D.G. Andersen, G.R. Ganger, G.A. Gibson, B. Mueller, Safe and effective fine-grained TCP retransmissions for datacenter communication, in *Proceedings of the ACM SIGCOMM 2009 Conference on Data Communication*, ser. SIGCOMM '09. New York: ACM, 2009, pp. 303–314. http://doi.acm.org/10.1145/1592568. 1592604

24. V. Vasudevan, A. Phanishayee, H. Shah, E. Krevat, D.G. Andersen, G. R. Ganger, G.A. Gibson, A (In)cast of thousands: scaling datacenter TCP to kiloservers and gigabits, Technical Report CMUPDL-09-101, Carnegie Mellon University Parallel Data Lab, Feb 2009. http://www.pdl. cmu.edu/PDL-FTP/Storage/CMU-PDL-09-101.pdf

25. Y. Chen, R. Griffith, J. Liu, R.H. Katz, A.D. Joseph, Understanding TCP incast throughput collapse in datacenter networks, in *Proceedings of the 1st ACM Workshop on Research on Enterprise Networking*, ser. WREN '09. New York: ACM, 2009, pp. 73–82. http://doi.acm. org/10.1145/1592681.1592693

26. G.A. Gibson, D.F. Nagle, K. Amiri, J. Butler, F.W. Chang, H. Gobioff, C. Hardin, E. Riedel, D. Rochberg, J. Zelenka, A cost-effective, high-bandwidth storage architecture, in *Proceedings of the Eighth International Conference on Architectural Support for Programming languages and Operating Systems*, ser. ASPLOS-VIII. New York: ACM, 1998, pp. 92–103. http://doi.acm. org/10.1145/291069.291029

27. D. Nagle, D. Serenyi, A. Matthews, The panasas active scale storage cluster: delivering scalable high bandwidth storage, in *Proceedings of the 2004 ACM/IEEE Conference on Supercomputing*, ser. SC '04. Washington, DC, USA: IEEE Computer Society, 2004, pp. 53-. http://dx.doi. org/10.1109/SC.2004.57

28. S. Ghemawat, H. Gobioff, S.-T. Leung, The Google file system, SIGOPS Oper. Syst. Rev 37(5), 29–43 (2003). http://doi.acm.org/10.1145/1165389.945450

29. M. Abd-El-Malek, W.V. Courtright, II, C. Cranor, G.R. Ganger, J. Hendricks, A.J. Klosterman, M. Mesnier, M. Prasad, B. Salmon, R.R. Sambasivan, S. Sinnamohideen, J.D. Strunk, E. Thereska, M. Wachs, J.J. Wylie, Ursa minor: versatile cluster-based storage, in *Proceedings of the 4th Conference on USENIX Conference on File and Storage Technologies - Volume 4*, ser. FAST'05. Berkeley: USENIX Association, 2005, pp. 5–5. http://dl.acm.org/citation.cfm? id=1251028.1251033

30. Y. Cheng, C. Qin, F. Rusu, GLADE: big data analytics made easy, in *Proceedings of the 2012 ACM SIGMOD International Conference on Management of Data*, ser. SIGMOD '12. New York: ACM, 2012, pp. 697–700. http://doi.acm.org/10.1145/2213836.2213936

31. K. Bajda-Pawlikowski, D.J. Abadi, A. Silberschatz, E. Paulson, Efficient processing of data warehousing queries in a split execution environment, in *Proceedings of the 2011 ACM SIGMOD International Conference on Management of Data*, ser. SIGMOD '11. New York: ACM, 2011, pp. 1165–1176. http://doi.acm.org/10.1145/1989323.1989447

32. Y. Huai, R. Lee, S. Zhang, C. H. Xia, X. Zhang, DOT: a matrix model for analyzing, optimizing and deploying software for big data analytics in distributed systems, in *Proceedings of the 2nd ACM Symposium on Cloud Computing*, ser. SOCC '11. New York: ACM, 2011, pp. 4:1–4:14. http://doi.acm.org/10.1145/2038916.2038920

33. F.J. Alexander, A. Hoisie, A.S. Szalay, Big data [Guest editorial]. Comput. Sci. Eng. 13(6), 10–13 (2011)

34. J. Dean, S. Ghemawat, MapReduce: simplified data processing on large clusters, Commun. ACM 51(1), 107–113 (2008). http://doi.acm.org/10.1145/1327452.1327492

35. K. Shvachko, H. Kuang, S. Radia, R. Chansler, The hadoop distributed file system, in *Proceedings of the 2010 IEEE 26th Symposium on Mass Storage Systems and Technologies (MSST)*, ser. MSST '10. Washington: IEEE Computer Society, 2010, pp. 1–10. http://dx.doi.org/10.1109/MSST.2010.5496972

36. J. Padhye, V. Firoiu, D. Towsley, J. Kurose, Modeling TCP throughput: a simple model and its empirical validation, in *Proceedings of the ACM SIGCOMM '98 Conference on Applications, Technologies, Architectures, and Protocols for Computer Communication*. New York: ACM, 1998, pp. 303–314. http://doi.acm.org/10.1145/285237.285291

37. N. Parvez, A. Mahanti, C. Williamson, An analytic throughput model for TCP NewReno, IEEE/ACM Trans. Netw 18(2), 448–461 (2010). http://dx.doi.org/10.1109/TNET.2009.2030889

38. E. Altman, K. Avrachenkov, C. Barakat, A stochastic model of TCP/IP with stationary random losses, IEEE/ACM Trans. Netw 13(2), 356–369 (2005). http://dx.doi.org/10.1109/TNET.2005.845536

39. M. Goyal, R. Gurin, R. Rajan, Predicting TCP throughput from non-invasive network sampling, in *I*NFOCOM, 2002

40. A. Kumar, Comparative performance analysis of versions of TCP in a local network with a lossy link, IEEE/ACM Trans. Netw 6(4), 485–498 (1998). http://dx.doi.org/10.1109/90.720921

41. C. Casetti, M. Meo, A new approach to model the stationary behavior of TCP connections, in *I*NFOCOM 2000. Nineteenth Annual Joint Conference of the IEEE Computer and Communications Societies. Proceedings. IEEE, vol. 1 (2000), pp. 367–375

42. S. Floyd, V. Jacobson, Random early detection gateways for congestion avoidance, IEEE/ACM Trans. Netw 1(4), 397–413 (1993). http://dx.doi.org/10.1109/90.251892

43. K. Fall, S. Floyd, Simulation-based comparisons of Tahoe, Reno and SACK TCP, SIGCOMM Comput. Commun. Rev 26(3), 5–21 (1996). http://doi.acm.org/10.1145/235160.235162

44. S. Mccanne, S. Floyd, K. Fall, ns2 (network simulator 2), http://www-nrg.ee.lbl.gov/ns/.http://www.isi.edu/nsnam/ns/.

45. K. Varadhan, K. Fall, The ns Manual (formerly ns Notes and Documentation), The VINT Project, A Collaboration between researchers at UC Berkeley, LBL, USC/ISI, and Xerox PARC, Nov 2011. http://www.isi.edu/nsnam/ns/doc/ns_doc.pdf

Chapter 4
Addressing TCP Incast

As discussed in Chaps. 1 and 3, clients performing synchronized reads across an increasing number of servers in high bandwidth, low latency data center environments, observe TCP's throughput drop by one or two orders of magnitude below their link capacity. Labeled Incast, this pathological behavior of TCP is endured by a growing number of data center applications and services. Hence, a feasible solution that addresses the Incast problem is urgently needed. In this chapter, we provide a broad overview of existing Incast solutions followed by detailed description of our proposed techniques that are designed to address the Incast problem at the Transport Layer [1].[1]

4.1 Existing Solutions

Since timeouts are the primary reason behind TCP Incast, in this section, we shall briefly discuss existing solutions that either avoid timeouts or reduce their penalty. While all the solutions discussed here are moderately effective in masking Incast, only two techniques discussed in Sects. 4.1.3 and 4.1.4, manage to accomplish this at the transport layer.

4.1.1 Larger Switch Buffers

This Incast solution, discussed in [2], tries to mitigate the root cause of timeouts—packet losses—by increasing the buffer space allocated per port on the Ethernet switches. To evaluate this solution, we vary the size of the switch port buffers in the cluster based storage system detailed in Sect. 3.2. Furthermore, to match the setup

---

[1] All simulations discussed in this chapter use the same topology as depicted in Fig. 3.7. Furthermore, unless noted explicitly, the simulations use the same parameters and values as listed in Table 3.1.

S. Kulkarni and P. Agrawal, *Analysis of TCP Performance in Data Center Networks*,   67
SpringerBriefs in Electrical and Computer Engineering,
DOI: 10.1007/978-1-4614-7861-4_4, © The Author(s) 2014

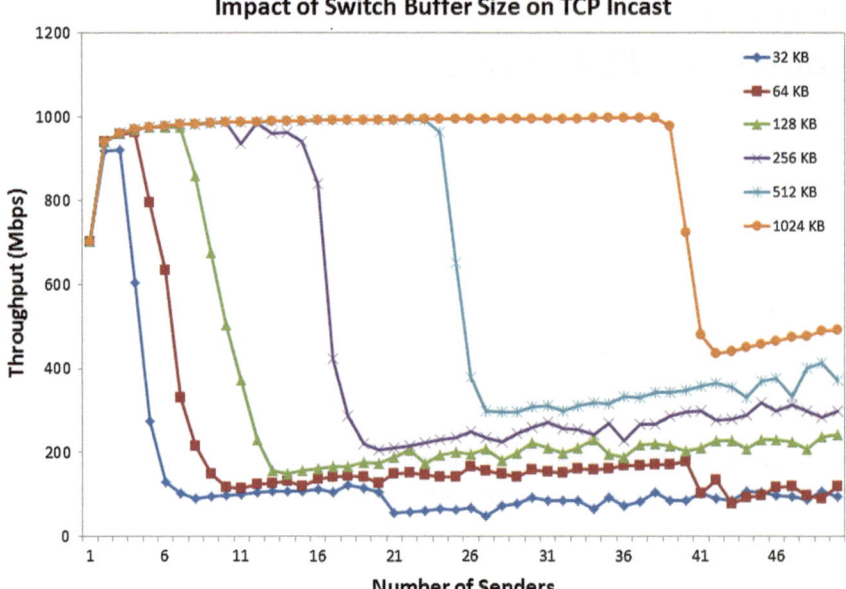

Fig. 4.1 Effect of the size of switch buffers on TCP Incast

presented in [2], we configure our network link delay to 100 μs and limit TCP's receive window size to 20 segments (the rest of the parameters remain as outlined in Table 3.1). The results of this experiment are depicted in Fig. 4.1. Please note that unlike the measure packets/sec used to discuss results in Chap. 3, throughput in Fig. 4.1 is measured in Mbps. Figure 4.1 clearly shows that doubling the size of the switch's output port buffer, doubles the number of servers that can supported before the onset of Incast.

Consequently, given the number of servers, Incast can be avoided with a large enough buffer space. Unfortunately, switches with larger buffers tend to cost a lot more, forcing system designers to choose between over-provisioning and hardware budgets. This suggests that a more cost-effective solution is needed to address TCP Incast.

4.1.2 Increasing SRU Size

This is another Incast countermeasure discussed in [2]. It aims to mask TCP's throughput collapse by utilizing the spare link capacity of the stalled flow in transferring larger SRUs belonging to other flows. To evaluate this solution, we vary the size of the SRUs in the cluster based storage system discussed in Sect. 3.2, while limiting the size of the switch port buffer to 32 KB. The results of this experiment

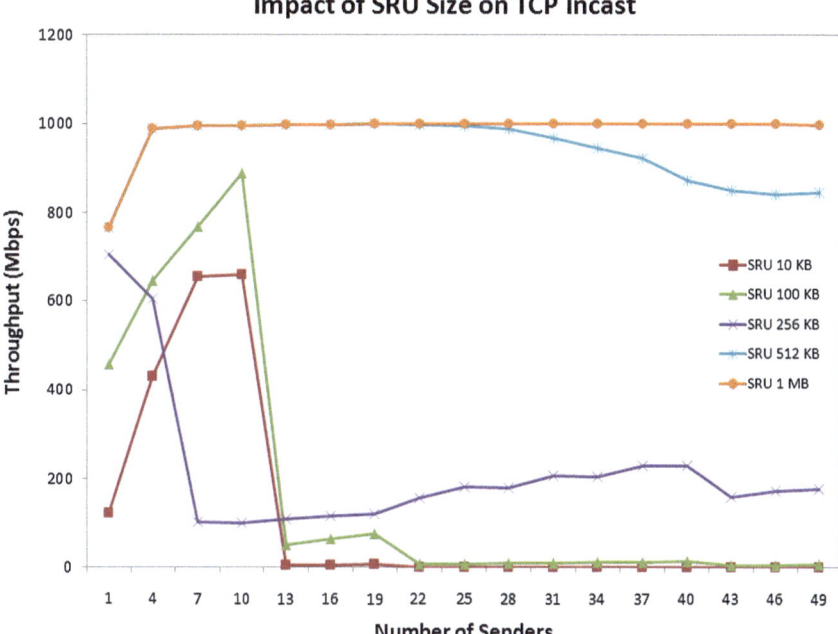

Fig. 4.2 Effect of the size of the SRUs on TCP Incast

are depicted in Fig. 4.2. Figure 4.2 illustrates that increasing the size of the SRUs, improves the overall throughput at the client. For example, with seven servers, the throughput for 1 MB SRU is orders of magnitude greater than the throughput of SRU of size 256 KB.

As discussed in Chap. 3, TCP performs well in settings without synchronized reads, which can be modeled by infinite sized SRUs. With large SRUs, the servers take longer to complete transmitting their share of data. This allows the active servers to utilize the spare link capacity made available by the stalled flows during timeouts. In doing so, the servers effectively reduce the idle link time experienced by the client, which in turn improves its overall throughput.

Unfortunately, SRU of size 1 MB is quite impractical; most applications ask for data in small chunks, corresponding to a size range of 1–256 KB. This is because, larger the size of the SRU, greater is the prefetching that the storage system has to commit to. With prefetching, the storage system needs to allocate pinned space in the client kernel memory, increasing the memory pressure at the client [3]. This increased pressure at the client, often leads to kernel failures. Hence it is really not advisable to use larger SRUs on cluster based storage systems.

4.1.3 Reducing Timeout Penalty

This technique, proposed in [3], aims to address TCP Incast by reducing the time spent in waiting for a timeout to end.

The amount of time a flow waits before retransmitting a lost packet without the duplicate ACK assisted Fast Retransmit mechanism, is determined by TCP's RTO value. Estimating TCP's RTO value involves achieving timely response to packet losses and also avoiding the occurrence of premature timeouts. Premature timeouts have the following negative effects:

- They lead to spurious retransmissions which can potentially cause and prolong network congestion.
- They cause TCP to enter the Slow Start recovery after reducing its Slow Start Threshold (*ssthresh*) value by half, even when no packets were lost. In doing this, the protocol underestimates its link capacity resulting in lower throughput for its users.

TCP therefore, has a conservative minimum RTO (RTO_{min}) value to guard itself against the ill effects of spurious retransmissions [4, 5].

Popular implementations of TCP use a RTO_{min} value of 200 ms [6]. Although this value is appropriate in wide area networks, it is orders of magnitude greater than the round trip times in data center networks. This large RTO_{min} value, imposes a huge penalty on TCP's throughput as the transfer times for segments within a data center, are significantly smaller than the value of its RTO_{min}.

In [3], the authors suggest reducing the value of RTO_{min} from 200 ms to 200 µs, in order to lessen the penalty of TCP timeouts on synchronized reads. To evaluate this solution, we decrease the value of TCP's RTO_{min} in the cluster based storage system discussed in Sect. 3.2, while limiting the size of the switch port buffer to 32 KB. The results of this experiment are depicted in Fig. 4.3. From Fig. 4.3, it is clear that reducing TCP's RTO_{min} value, improves the overall throughput at the client even after taking into account, the drop in peak performance when the number of servers is greater than 40.

In general, for any given SRU size, reducing RTO_{min} value improves the overall throughput at the client. Unfortunately, setting RTO_{min} to 200 µs poses the following challenges:

- According to RTO computing algorithms in [4, 5], reducing RTO_{min} to 200 µs requires a TCP clock granularity of 100 µs. TCP implementations on most operating systems including the likes of BSD and Linux, are currently unable to provide this fine grained timer. For example, BSD implementation of TCP, expects the operating system to provide two coarse-grained "heartbeat" software interrupts every 200 ms and 500 ms, which are used to handle internal per-connection timers [7]. Similarly, TCP implementation on Linux, expects a clock granularity of 10 ms

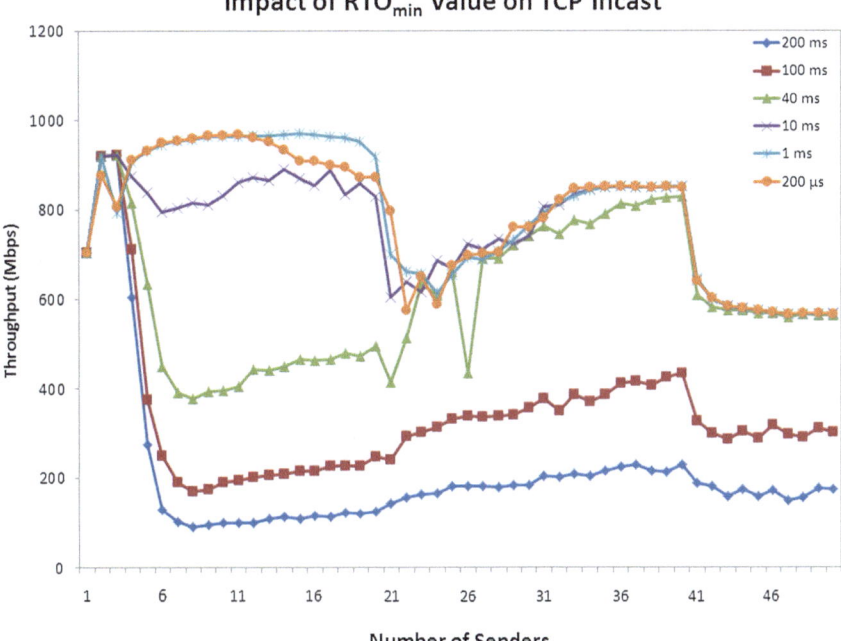

Fig. 4.3 Effect of the RTO$_{min}$ value on TCP Incast

from the operating system. Some operating systems can support fine grained timers by either employing specialized external hardware or utilizing high resolution software timers [8]. However, neither of these options are feasible in the context of data centers. External hardware scales poorly inside a data center while software timers which require kernel changes, are not supported by all operating systems.

- Even if sufficiently fine grained TCP timers were supported, reducing RTO$_{min}$ value can be harmful, especially in situations where the servers communicate with clients outside the data center. In [9], the authors note that low values for RTO$_{min}$ increases the occurrence of premature timeouts as RTO$_{min}$ can be used for trading "timely response with premature timeouts". Other studies of RTO estimation in similar high-bandwidth, low-latency ATM networks also show that very low RTO$_{min}$ values result in spurious retransmissions [10] because variations in round-trip-times inside wide-area networks clash with the standard RTO estimator's short RTT memory.

In summary, the solution proposed in [3] should be viewed with caution as it increases the risk of premature timeouts.

4.1.4 Relying on Explicit Congestion Notification

Data Center TCP (DCTCP), is a protocol proposed in [11]. It aims to achieve high burst tolerance, low latency and high throughput during synchronized data transfers, by requiring Ethernet switches to support Explicit Congestion Notifications (ECN).

DCTCP relies on a simple marking scheme at switches that sets the Congestion Experienced (CE) codepoint of packets as soon as the buffer occupancy exceeds a fixed small threshold. DCTCP uses these ECNs to provide multi-bit feedback to its end hosts. The DCTCP source reacts to such notifications by reducing the window by a factor that depends on the fraction of marked packets: larger the fraction, bigger is the decrease factor.

Unfortunately, not all switches support ECN. Without the underlying ECN support, DCTCP faces the same issues and hurdles as standard TCP. Additionally, ECNs are known to be effective in simple configurations only. With more than one switch, ECNs have an adverse effect on data flows [2]. Furthermore, authors in [11], make no claims about he suitability of DCTCP for wide area networks as they assume internal data center traffic to be separate from that of the external world.

4.2 Probabilistic Retransmission

In TCP world, timeouts are indicators of severe network congestion. Although the penalty for detecting congestion through timeouts is quite large in TCP, they are unavoidable in certain scenarios like, full window losses and retransmission losses. In this section, we shall examine a technique that reduces the time taken in detecting network congestion when TCP's loss recovery mechanism cannot be triggered by duplicate ACKs. Specifically, we shall explore the notion of proactively detecting network congestion through probabilistic retransmissions, while using TCP's retransmission timer as a fallback option.

4.2.1 Retransmit Thread

As discussed in Sect. 4.1.3, TCP has a conservative minimum RTO (RTO_{min}), whose value is orders of magnitude greater than the round trip times at data centers. To overcome the penalty imposed by a conservative RTO_{min} on timeouts in synchronized workloads, we propose a congestion recovery technique that relies on probabilistic retransmissions, kernel threads and duplicate ACKs.

Most modern operating systems support threads in their kernel space. A kernel thread is the "lightest" unit of kernel scheduling. Our solution to the Incast problem utilizes one such kernel thread to probabilistically retransmit the *highest unacknowl-edged segment* in sender's transmission window. That is, every time the thread is scheduled for execution, it retransmits with probability p, the *highest unacknowledged segment* in sender's transmission window. Before retransmitting the segment, the thread also "marks" it as being *'probabilistically retransmitted'*. Algorithm 1 captures necessary details regarding the Retransmit Thread.

Algorithm 1 Retransmit Thread at Sender

---

if $length(Transmit\ Window) \geq 1$ **then**
  if $uniform(0, 1) \leq p$ **then**
    mark $Highest\ UnACKed\ Segment$
    retransmit $marked\ Segment$
  end if
end if
yield $processor$

To "mark" the segment as being *'probabilistically retransmitted'*, the Retransmit Thread uses one of the six reserved bits in the segment's header. Figure 2.1 shows the layout of a TCP segment with the reserved bits located next to the Header Length field.

Because of its probabilistic nature, the retransmitted segment can arrive at the Ethernet switch (i) before any congestion, (ii) during a congestion or (iii) after a congestion. Case (i) would result in the destination receiving multiple copies of the same segment—the original segment transmitted by TCP, followed by the "marked" segment transmitted by our Retransmission Thread. In this situation, the client ignores the "mark" on the retransmitted segment and responds back with a normal cumulative ACK. In case (ii), the retransmitted segment is dropped by the switch since it arrives at a time when the switch's port buffers are full. Since the "marked" segment never reaches the destination, neither the sender nor the receiver are required to take any action. Under case (iii), if the sender's original segment was dropped at the switch due to congestion, the receiver would be seeing the sequence number on the retransmitted segment for the first time. Since the first copy of the segment is itself "marked", the receiver responds back with a normal cumulative ACK followed by three duplicate ACKs. By doing this, not only does the receiver acknowledge the occurrence of a congestion at the intermediate switch, but it also helps the sender trigger Fast Retransmit for quicker loss recovery. Algorithm 2 lists the steps involved in handling retransmitted segments at the receiver.

When the sender receives three duplicate ACKs in a row, it automatically performs loss recovery using Fast Retransmit mechanism, without waiting for retransmission timer to expire. Algorithm 3 gives details on handling duplicate ACKs at the sender.

Receiving a "marked" segment with an unseen sequence number indicates that (i) there was congestion in the network which accounted for the original copy of the segment, and (ii) the congestion is now cleared, for the "marked" segment would never have made it through otherwise. With congestion in the network now resolved, the receiver would like the sender to start its loss recovery early, without having to wait for a retransmission timer to expire. It initiates this by sending three duplicate

Algorithm 2 Handling Retransmitted Segments at Receiver

...normal handling of segment...
send *ACK*
if *isduplicate(ReceivedSegment)* ≡ *false* **then**
  if *ismarked(ReceivedSegment)* ≡ *true* **then**
    for *i* = 1 to 3 **do**
      send *ACK*
    end for
  end if
end if

Algorithm 3 Handling ACKs at Sender

...normal handling of ACK...
if *dupackcount* ≡ 3 **then**
  suspend *retransmission thread*
  invoke *Fast Retransmit*
end if

ACKs back to the sender which forces the sender to immediately perform an smooth reduction of its flow via Fast Recovery, instead of performing an abrupt reduction through Slow Start following a timeout.

It is also possible that our Retransmission Thread never retransmits the highest unacknowledged segment. In such a case, the sender detects and responds to congestion only when its retransmission timer expires.

4.2.2 Performance Analysis

In order to measure the effectiveness of the suggested technique, we implement Algorithms 1, 2 and 3 in ns-2. To keep the simulations realistic, we model the thread context switch time by including a small delay of $20\,\mu s$ between each execution of the Retransmission Thread. We also fix the RTO_{min} value to 200 ms. The rest of the experimental setup is the same as the one described in Sect. 4.1.3. Figure 4.4 shows that increasing the value of p (probability of retransmission), improves the throughput at the client by orders of magnitude, when the number of senders is greater than eight.

From Fig. 4.4, it is clear that using Retransmission Threads can significantly improve TCP's performance under synchronized workloads. However, the value of its retransmission probability, p, should be chosen with some consideration. If p is set too low, the proposed technique provides no significant benefits over default TCP. On the other hand, if p is set too high, it causes unnecessary retransmissions, contributing further to the congestion at the switch. Figure 4.5 shows the drop ratio i.e., the number of packets dropped at the switch versus the number of packets received by it, for varying values of p. The graph also includes plots for default TCP with RTO 200 ms as well as modified TCP with RTO $200\,\mu s$, for reference. For optimal

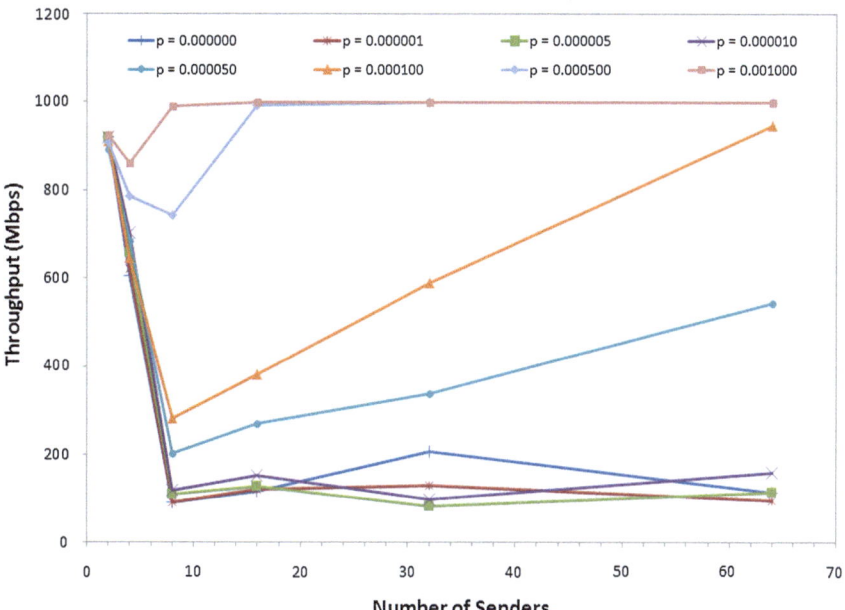

Fig. 4.4 Effect of retransmission probability, p, on TCP Incast

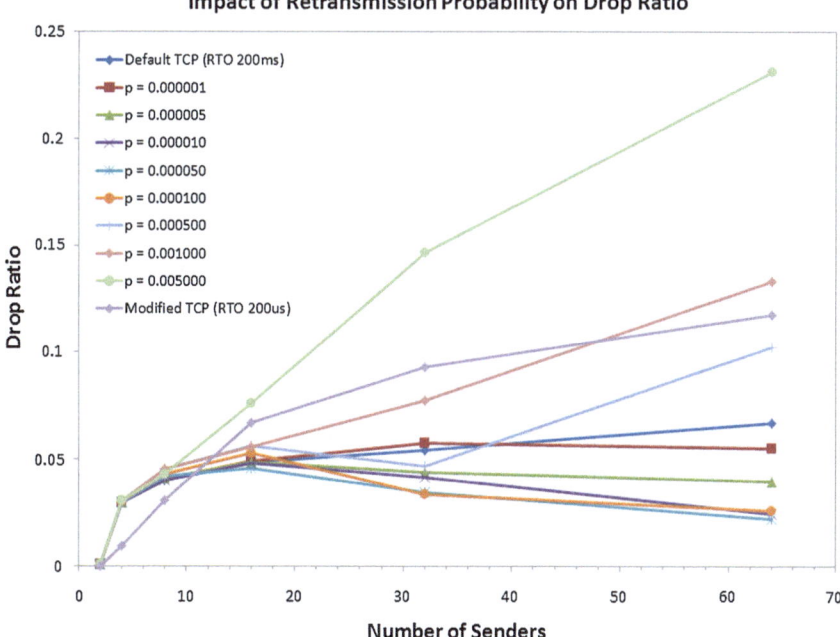

Fig. 4.5 Effect of retransmission probability, p, on drop ratio

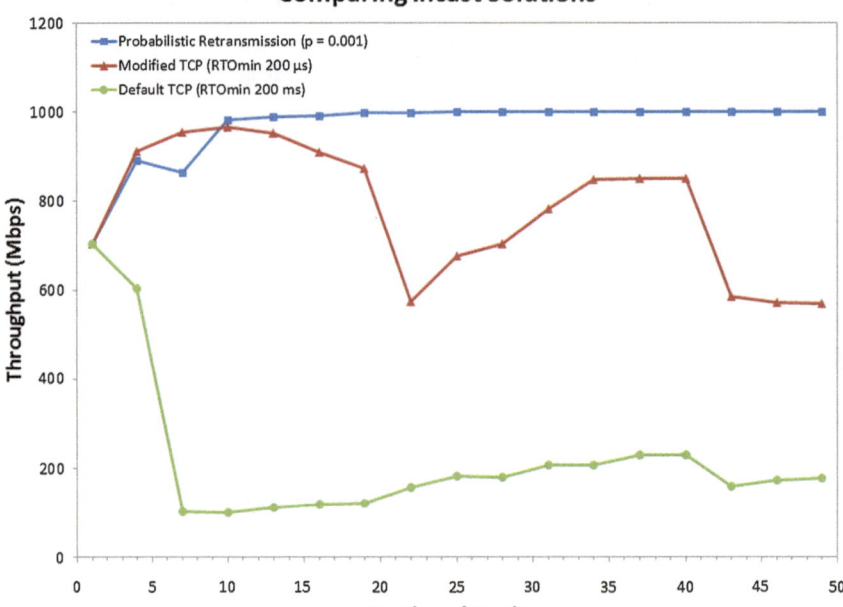

Fig. 4.6 Comparing probabilistic retransmission with default and modified TCP

p, the probabilistic retransmission technique would yield high TCP throughput with a low drop ratio. From Figs. 4.4 and 4.5, it is clear that for our simulation environment, the best value of *p* is 0.001.

Figure 4.6 compares the performance of probabilistic retransmission ($p = 0.001$) with default TCP (RTO 200 ms) and modified TCP (RTO 200 μs). From Fig. 4.6, it is evident that the probabilistic retransmission outperforms default TCP under all experimental conditions. The technique also performs better than the modified TCP, when the number of senders in the experiment is greater than ten. On the other hand, when the number of senders in the experiment is between five and ten, modified TCP yields slightly better throughput than our proposed solution. This is because, very few senders experience severe losses when the sender count in the experiment is less than ten. In addition to that, the value of the retransmission probability, *p*, is only 0.001. Therefore, it is quite likely that the loss experiencing senders make several attempts before succeeding at their probabilistic retransmissions. This in turn leaves the switch-client link underutilized for some period which results in a small dip in the solution's performance when compared to modified TCP.

However, when the number of loss experiencing senders is large, it is more likely that at least one of them will quickly succeed in its probabilistic retransmission. With every such success, the switch-client link is kept occupied for that much longer, resulting in a performance that is significantly better than that of the modified TCP.

One must also keep in mind that the results discussed above are true only for the chosen value for p, in this case 0.001. In Fig. 4.4, we saw that higher values of p need fewer senders to achieve throughput saturation. Hence, if the synchronized workload inside a data center involves only a few senders, probabilistic retransmission can still outperform modified TCP, if p is set to a higher value.

4.2.3 Summary

Based on our experiments and analysis, it is clear that probabilistic retransmission offers a feasible solution to TCP's Incast problem. In addition to being backwards compatible with existing flavors of TCP, the technique is also able to outperform existing Incast solutions, without incurring any of their drawbacks.

However, probabilistic retransmission relies heavily on the availability of kernel threads. Also, its performance is governed by the value assigned to p, the retransmission probability. Ideally, the value of p should be auto computed and auto tuned, but we take the easier option for now, and make it a user configurable variable. As part of our future work, we plan to implement this technique on a Linux based cluster and measure its performance in the real world.

4.3 Dynamic Segment Resizing

As detailed in Chap. 2, when TCP receives an out-of-order segment, it immediately responds back with a duplicate ACK. From the sender's perspective, receiving a duplicate ACK indicates potential loss or reordering of transmitted segments. TCP's Fast Retransmit algorithm uses the arrival of three consecutive duplicate ACKs as an indication that segments have been lost. The algorithm then initiates loss recovery at the sender, without waiting for the retransmission timer to expire. But, when the destination receives fewer than four segments due to severe network congestion, it has no chance of sending three duplicate ACKs, meaning, retransmission timeouts are the only means of loss recovery for a source, that has lost all its segments to network congestion.

Timeouts are known to have a negative impact on TCP's performance since, the time needed for the protocol to recover losses through retransmission timer is much longer than the time needed to recover via Fast Retransmit algorithm. As discussed in Chaps. 1 and 3, timeouts are also known to cause the Incast problem that TCP experiences during synchronized data transfers. In our proposed scheme, we aim to address TCP Incast by making loss recovery through Fast Retransmit possible in operating regions where currently, timeouts are the only option available.

Dynamic Segment Resizing is based on the idea of increasing the upstream flow of ACKs by sending downstream, a large number of segments whose size is smaller than the maximum segment size supported by the connection. When a large number

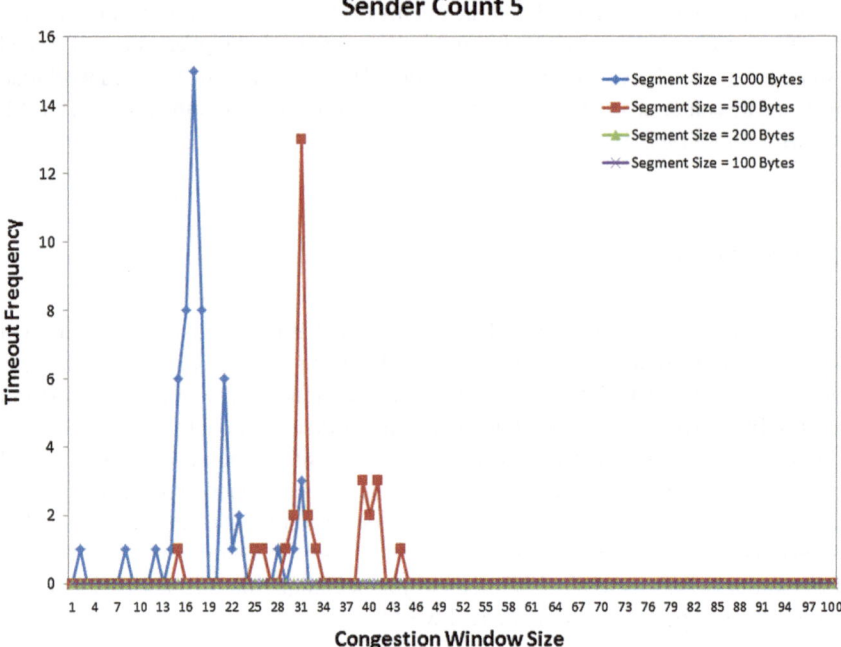

Fig. 4.7 Timeout frequency for different segment sizes when sender count is 5

of segments are received at the destination, it triggers a large number of ACKs on the back channel. And, larger the number of ACKs on the back channel, larger is the probability of the source recovering lost segments without the aid of a retransmission timer. In other words, our proposed procedure gives the transmitter a chance to obtain more information about the current state of the network between itself and the receiver.

To illustrate our approach by means of an example, we vary the size of TCP's segments in the cluster file system experiment discussed in Sect. 3.2. In this experiment, we also limit the port buffer length on the intermediate switch to 32 KB, set the size of the SRU to 256 KB, cap the receive window size to 32 KB and fix the value of the minimum retransmission timeout, RTO_{min}, to 200 ms.

Figures 4.7, 4.8, 4.9 and 4.10, depict the effects of smaller TCP segments on the protocol's retransmission timeouts when the number of senders in the experiment is 5, 10, 20 and 50 respectively. From these figures, it is clear that smaller sized segments reduce the number of timeouts that TCP experiences during a synchronized transfer. Additionally, smaller sized segments move the peak of the timeout histogram to the right, meaning, with smaller segments, TCP will have to lose a greater number of packets to experience a timeout. The graphs also suggest that with small enough segments, TCP can completely avoid timeouts during synchronized data transfers.

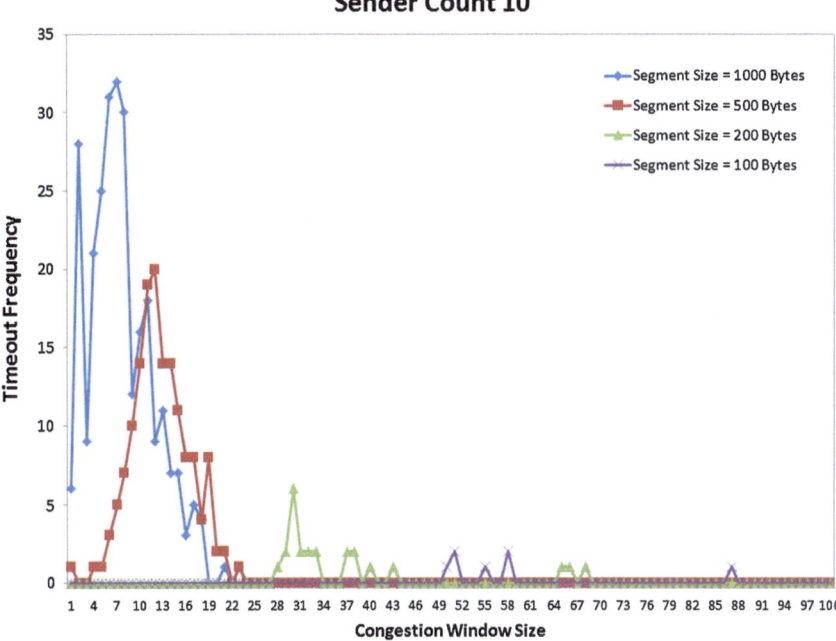

Fig. 4.8 Timeout frequency for different segment sizes when sender count is 10

Apart from the aforementioned advantages, reducing the size of the segments also gives TCP a finer control over the amount of data that can be outstanding in the network. However, transmitting smaller sized segments decreases TCP's line efficiency, which is defined as the ratio of data size to the size of (header + data) in a segment. In order to improve TCP's line efficiency when operating with smaller sized segments, we employ a header compression technique that is described in [12]. This data compression mechanism, reduces the normal 40 byte TCP/IP packet headers down to 3–4 bytes in average case. It does this by saving the state of TCP connections at both ends of a link, and only sending the differences in the header fields that change. With this header compression technique in place, even a small data segment of 36 bytes will be able to achieve a line efficiency of 90 % for TCP.

In Figs. 4.7, 4.8, 4.9 and 4.10, we notice that different cluster configurations have different limits on segment sizes that allow synchronized transfers to take place without incurring any timeout penalty. In order to maximize TCP's line efficiency during synchronized transfers involving smaller segments, it is desirable to have segment sizes that operate closer to these limits. Dynamic Segment Resizing is able to achieve this by relying on a congestion window threshold value called $cwnd_{dsr}$. The solution mandates TCP to begin its synchronized transfer with a predefined segment size of MSS_{dsr} bytes. As TCP starts transmitting user data, its congestion window begins to grow. When TCP's congestion window, $cwnd$, grows beyond the congestion

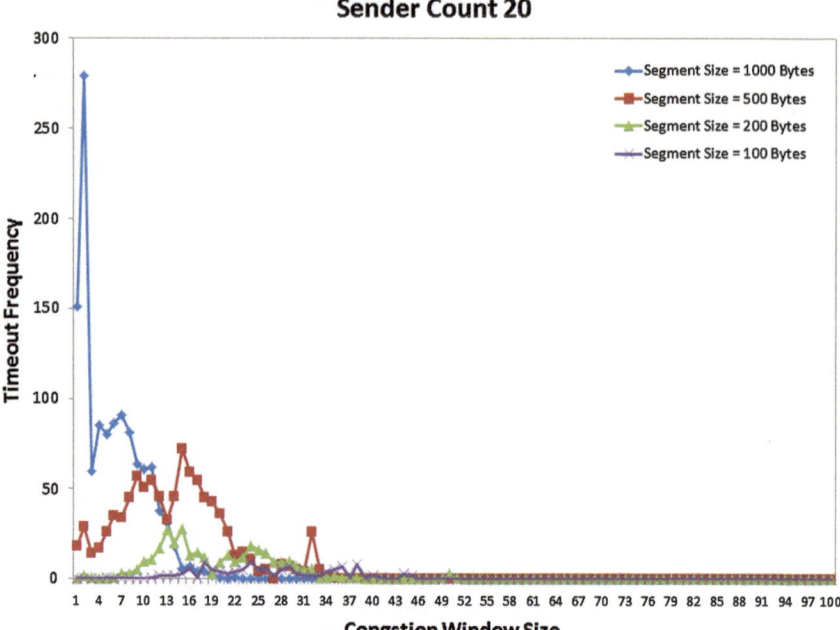

Fig. 4.9 Timeout frequency for different segment sizes when sender count is 20

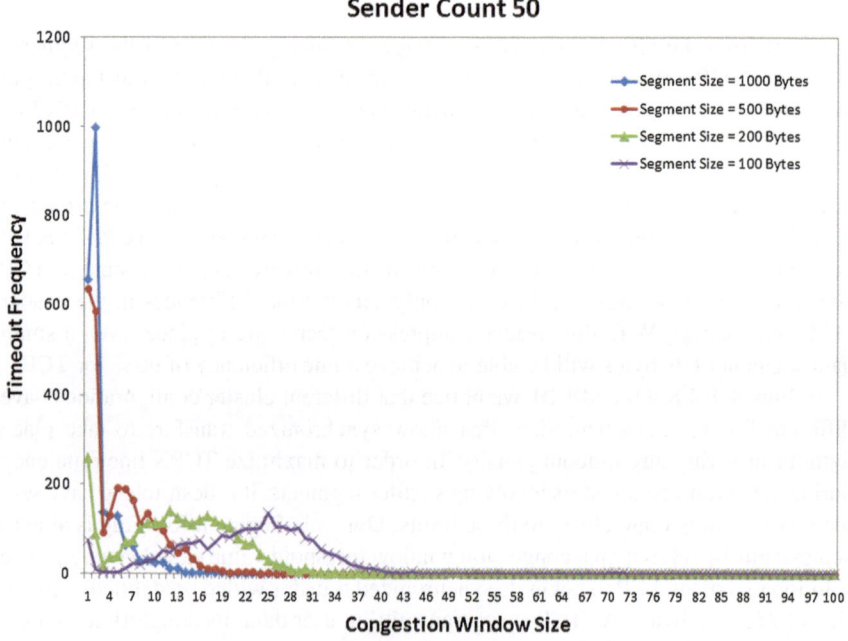

Fig. 4.10 Timeout frequency for different segment sizes when sender count is 50

window threshold, $cwnd_{dsr}$, our solution resizes TCP's segments to $\frac{(cwnd \times MSS_{dsr})}{\left(\frac{cwnd_{dsr}}{2}\right)}$ bytes. TCP's congestion window, $cwnd$, is also resized to $\left(\frac{cwnd_{dsr}}{2}\right)$ segments. Following this resize procedure, TCP resumes transmitting user data albeit with slightly bigger segments. As before, the segments are resized if TCP's $cwnd$ again grows beyond $cwnd_{dsr}$. This resize-transmit-resize cycle continues as long as TCP's segments remain smaller than the maximum segment size of the connection and its congestion window, $cwnd$, continues to grow beyond the threshold, $cwnd_{dsr}$. The cycle is eventually broken when the size of the resized segments equal the MSS of the connection or when the flow encounters duplicate ACKs which prevent the congestion window from growing beyond $cwnd_{dsr}$. Algorithm 4 captures the necessary details regarding the resize procedure.

Algorithm 4 Resize Procedure for Dynamic Segment Resizing

...normal handling of cwnd growth...
if $(MSS_{dsr} < MSS)$ **and** $(cwnd \geq cwnd_{dsr})$ **then**
  $MSS_{temp} = \frac{(cwnd \times MSS_{dsr})}{\left(\frac{cwnd_{dsr}}{2}\right)}$
  if $MSS_{temp} > MSS$ **then**
    $MSS_{temp} = MSS$
  end if
  $cwnd = \frac{(MSS_{dsr} \times cwnd)}{MSS_{temp}}$
  $MSS_{dsr} = MSS_{temp}$
end if

Dynamic Segment Resize is a proposed Incast solution that requires some minor changes to the sender's TCP stack. These changes are easy to incorporate and only require a few modifications to existing TCP code. The header compression procedure on the other hand, needs to be implemented at both the communicating endpoints.

4.3.1 Performance Analysis

In order to measure the effectiveness of the suggested technique, we implement Algorithm 4 in ns-2. We then measure the performance of Dynamic Segment Resizing technique in the cluster file system example discussed in Sect. 3.2. For this experiment, we limit the port buffer size on the intermediate switch to 32 KB, set the size of the SRU to 256 KB, fix the value of the minimum retransmission timeout, RTO_{min} to 200 ms and cap the receive window size at 1,000 segments. We also set Algorithm 4 specific variables, MSS_{dsr} and $cwnd_{dsr}$, to be 50 bytes and 50 segments respectively. The rest of the experimental setup is the same as the one described in Sect. 4.1.3.

Figure 4.11, compares the performance of Dynamic Segment Resizing with default TCP. From Fig. 4.11, it is evident that Dynamic Segment Resizing incurs a small penalty in performance when the number of senders in the experiment is

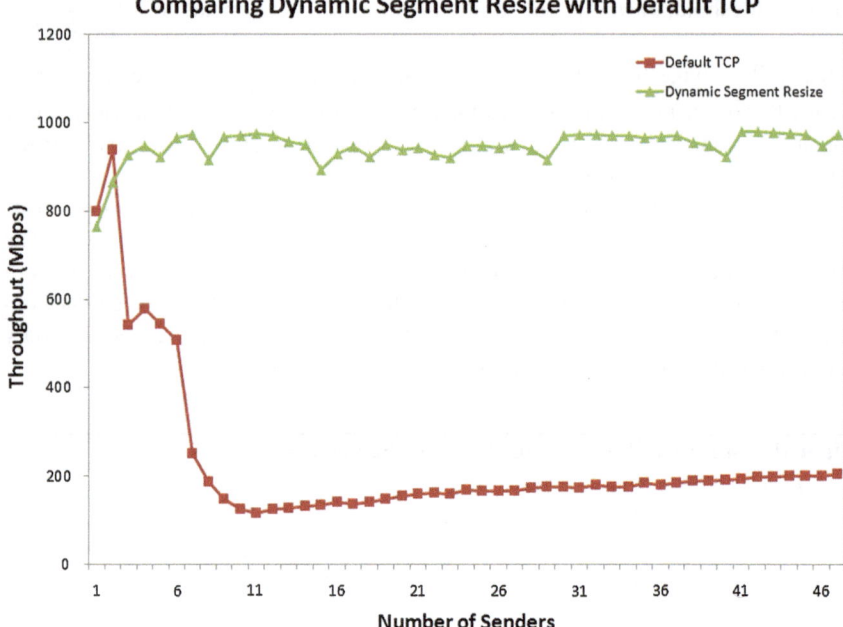

Fig. 4.11 Comparing dynamic segment resize with default TCP

less than three. This is because, the proposed technique takes some time to con-
verge on the connection's maximum segment size as the most appropriate size to
perform synchronized data transfers without incurring any timeout penalty. Default
TCP on the other hand, starts with maximum sized segments and therefore, is able to
achieve better line rate than Dynamic Segment Resizing. However, when the number
of senders in the cluster file system is greater than three, our proposed solution easily
outperforms default TCP.

4.3.2 Summary

From the simulation results discussed in Sect. 4.3.1, it is clear that Dynamic Segment Resizing offers a practical, transport-layer solution to the Incast problem. The
technique only requires some minor modifications on the sender side TCP and is
backwards compatible with many existing flavors of the protocol.

Unlike the Probabilistic Retransmission technique discussed in Sect. 4.2, Dynamic
Segment Resizing does not depend on the availability of external resources like
kernel threads. However, like p in Probabilistic Retransmission, the performance of
Dynamic Segment Resizing is also dependent on the initial values of MSS_{dsr} and
$cwnd_{dsr}$. Ideally, the start values of MSS_{dsr} and $cwnd_{dsr}$ are auto computed, but

we for now, make these variables user configurable. As part of our future work, we plan to implement Dynamic Segment Resizing on Linux cluster and measure its performance in the real world.

References

1. L.J. Miller, The ISO reference model of open systems interconnection: a first tutorial, in *Proceedings of the ACM '81 Conference, Services ACM '81*. (ACM, New York, 1981), pp. 283–288, http://doi.acm.org/10.1145/800175.809901
2. A. Phanishayee, E. Krevat, V. Vasudevan, D.G. Andersen, G.R. Ganger, G.A. Gibson, S. Seshan, Measurement and analysis of TCP throughput collapse in cluster-based storage systems, in *Proceedings of the 6th USENIX Conference on File and Storage Technologies, Services FAST'08*. (USENIX Association, Berkeley, 2008), pp. 12:1–12:14, http://dl.acm.org/citation.cfm?id= 1364813.1364825
3. V. Vasudevan, A. Phanishayee, H. Shah, E. Krevat, D.G. Andersen, G.R. Ganger, G.A. Gibson, B. Mueller, Safe and effective fine-grained TCP retransmissions for datacenter communication, in *Proceedings of the ACM SIGCOMM 2009 Conference on Data Communication, Services SIGCOMM '09*. (ACM, New York, 2009), pp. 303–314, http://doi.acm.org/10.1145/1592568. 1592604
4. V. Jacobson, Congestion avoidance and control. SIGCOMM Comput. Commun. Rev. **25**(1), 157–187, (1995), http://doi.acm.org/10.1145/205447.205462
5. V. Paxson, M. Allman, Computing TCP's retransmission timer, RFC 2988 (Proposed Standard), Internet Engineering Task Force, Nov. 2000, obsoleted by RFC 6298. http://www.ietf.org/rfc/ rfc2988.txt
6. P. Sarolahti, A. Kuznetsov, Congestion control in Linux TCP, in *Proceedings of the FREENIX Track: 2002 USENIX Annual Technical Conference*. (USENIX Association, Berkeley, 2002), pp. 49–62, http://dl.acm.org/citation.cfm?id=647056.715932
7. M. Aron, P. Druschel, TCP implementation enhancements for improving webserver performance, Tech. Rep. TR99-335, Rice University, 1999, http://citeseerx.ist.psu.edu/viewdoc/ summary?doi=10.1.1.9.5152
8. M. Aron, P. Druschel, Soft timers: efficient microsecond software timer support for network processing, ACM Trans. Comput. Syst. **18**(3), 197–228 (2000), http://citeseer.ist.psu. edu/aron99soft.html
9. M. Allman, V. Paxson, On estimating end-to-end network path properties, SIGCOMM Comput. Commun. Rev. **31**(2 Supp.) 124–151 (2001), http://doi.acm.org/10.1145/844193.844203
10. A. Romanow, S. Floyd, Dynamics of TCP traffic over ATM networks, in *Proceedings of the Conference on Communications Architectures, Protocols and Applications, ser. SIGCOMM '94*. (ACM, New York, 1994), pp. 79–88, http://doi.acm.org/10.1145/190314.190322
11. M. Alizadeh, A. Greenberg, D.A. Maltz, J. Padhye, P. Patel, B. Prabhakar, S. Sengupta, M. Sridharan, Data center TCP (DCTCP), in *Proceedings of the ACM SIGCOMM 2010 Conference, Services SIGCOMM '10*. (ACM, New York, 2010), pp. 63–74, http://doi.acm.org/10. 1145/1851182.1851192
12. V. Jacobson, Compressing TCP/IP headers for low-speed serial links, RFC 1144 (Proposed Standard), Internet Engineering Task Force, Feb. 1990, http://www.ietf.org/rfc/rfc1144.txt

Chapter 5
Conclusions and Future Work

In this chapter, we summarize the research discussed in this manuscript and follow it up with the directions for future work.

5.1 Summary of Research

In this manuscript, we studied TCP's performance under many-to-one synchronized traffic, when operating in high speed, low latency data center networks. In particular, we discussed the problem of TCP Incast, which causes the protocol's throughput to drop to almost a tenth of its link's available capacity. We derived an analytical model to investigate Incast and attributed TCP's throughput collapse to its timeouts. We also proposed some transport layer techniques to overcome Incast and evaluated their merits using ns-2 simulations.

In Chap. 1, we discussed Cloud Computing and its different components. We outlined how growing adoption of Cloud Computing is prompting service providers to spawn more data centers. We also discussed the cost and compatibility reasons that persuade service providers to employ Ethernet as the baseline communication fabric for their data centers. We then introduced the problem of TCP Incast that results from utilizing TCP in an environment where many of its assumptions are violated. In particular, we saw how TCP's throughput collapses catastrophically under many-to-one synchronized traffic, when operating in Ethernet-based, high speed, low latency data center networks.

In Chap. 2, we presented details on mechanisms that are responsible for TCP's reliable data transfer, flow control and congestion control. Our work in this chapter, provided the necessary background for Chaps. 3 and 4, where we have considered the problem of TCP Incast in greater detail.

In Chap. 3, we presented a simple model for TCP Incast. The model captures the essence of many-to-one synchronized workloads and expresses throughput as a function of packet loss probability. In particular, it takes into account the behavior

S. Kulkarni and P. Agrawal, *Analysis of TCP Performance in Data Center Networks*, 85
SpringerBriefs in Electrical and Computer Engineering,
DOI: 10.1007/978-1-4614-7861-4_5, © The Author(s) 2014

of multiple TCP flows in presence of loss induced duplicate acknowledgments and retransmission timeouts. The model yields a simple, closed form formula for calculating throughput of many-to-one synchronized traffic and attributes TCP's throughput collapse to two types of timeouts, ABTT and IBTT. We validated the model through extensive simulations done using ns-2 simulator. We found that our model provides a very good match to the observed Incast behavior. The formula resulting from our model, can be used for many purposes such as fast evaluation of Incast behavior and design of Incast free transport protocols.

In Chap. 4, we discussed few existing Incast solutions and their drawbacks. We then proposed two feasible solutions that addressed TCP Incast at the transport layer. Specifically, we developed solutions that improved TCP's performance under synchronized workloads by either proactively detecting network congestion through probabilistic retransmission or by dynamically resizing TCP's segments in order to avoid incurring timeout penalty. We also implemented these solutions in TCP and tested them extensively using ns-2 simulator. We found that our proposed solutions are both able to avoid timeouts and overcome the ill effects of throughput collapse during synchronized data transfers in high speed, low latency, data center environments.

5.2 Future Work

There are several lines of research arising from the work presented in this manuscript. Some research lines that should be pursued in the future include:

- *Accounting window limitation in Incast model*—The model presented in Chap. 3, does not consider the impact of window limitation per composite flow. At the beginning of TCP flow establishment, the receiver advertises a maximum buffer size which determines the maximum congestion window size $W_{f_{\max}}$. As a consequence, during a period without loss indications, the window size can grow up to $W_{f_{\max}}$, but will not grow beyond this value. Our Incast model must be tweaked to account for this scenario.
- *Accounting flavor specific features*—TCP New-Reno and TCP SACK are some of the most dominant flavors of TCP that are currently deployed in data center networks. In order to accurately model these protocols, we need to modify our Incast model presented in Chap. 3 to accommodate flavor specific features.
- *Developing techniques for loss rate estimation*—For empirical validation of our Incast model in Chap. 3, we estimated the loss rate probability based on the traces generated by our ns-2 simulator. Since traces are not always available, we need to understand and evaluate various techniques that help us in loss rate estimation.
- *Apply Markovian analysis*—The Incast model presented in Chap. 3, is very simple and less accurate. Markovian analysis on the other hand, is known to be detailed and precise. To better analyze the Incast phenomenon, we need to model Incast using Markovian models.

- *Timeout type based solution*—Neither the existing techniques nor our proposed solutions differentiate between the type of timeouts causing the Incast. It should be possible to design a solution that takes the type of timeouts, ABTT or IBTT, into consideration.
- *Auto computation of control variables*—We currently use statically selected values for our solution specific variables like p, MSS_{dr} and $cwnd_{dr}$. More work is needed to investigate means of automatically updating these variables in order to guarantee better Incast performance.
- *Implement in real world*—Since, almost all the results presented in this manuscript are based on ns-2 simulations, we need to check if our proposed solutions work well in the real world. Towards this end we need to implement the techniques of Probabilistic Retransmissions as well as Dynamic Segment Resizing on a Linux based cluster and measure their performance in the real world.